ENVIRONMENTAL IMPLICATIONS
of
EXPANDED COAL UTILIZATION

Related Pergamon Titles of Interest

Books

ASHLEY
Energy and the Environment – A Risk Benefit Approach

ENVIRONMENTAL RESOURCES LTD
Environmental Impact of Energy Strategies Within the EEC

MAROIS
Towards a Plan of Action for Mankind
Volume 1: Long Range Mineral Resources and Growth
Volume 2: Long Range Energetic Resources and Growth

ROYSTON
Pollution Prevention Pays

SIMEONS
Coal (Its Role in Tomorrow's Technology)

UNCRET
Recent Advances in Mining and Processing of Low-Grade and Submarginal Mineral Deposits

UNECE
Coal 1985 and Beyond

UNECE
Environmental Impact Assessment

UNECE
Environment and Energy

UNECE
Oils and Gases from Coal

WILLIAMS
Carbon Dioxide, Climate and Society

Journals

Environment International
Progress in Energy & Combustion Science

Full Details of All Pergamon Publications/Free Specimen of any Pergamon Journal Available on Request From Your Nearest Pergamon Office.

ENVIRONMENTAL IMPLICATIONS

of

EXPANDED COAL UTILIZATION

A Study by:
The Beijer Institute
The United Nations Environment Programme
The U.S.S.R. Academy of Sciences

Editorial Writers:

M.J. CHADWICK

and

NILS LINDMAN

Published for the Beijer Institute,
UNEP and the U.S.S.R. Academy of Sciences

PERGAMON PRESS

OXFORD · NEW YORK · TORONTO · SYDNEY · PARIS · FRANKFURT

U.K.	Pergamon Press Ltd, Headington Hill Hall, Oxford OX3 0BW, England
U.S.A.	Pergamon Press Inc., Maxwell House, Fairview Park, Elmsford, New York 10523, U.S.A.
CANADA	Pergamon Press Canada Ltd, Suite 104, 150 Consumers Road, Willowdale, Ontario M2J 1P9, Canada
AUSTRALIA	Pergamon Press (Aust) Pty. Ltd, PO Box 544, Potts Point, N.S.W. 2011, Australia
FRANCE	Pergamon Press SARL, 24 rue des Ecoles, 75240 Paris, Cedex 05, France
FEDERAL REPUBLIC OF GERMANY	Pergamon Press GmbH, Hammerweg 6, D-6242 Kronberg-Taunus, Federal Republic of Germany

First edition 1982
Reprinted 1983
Reprinted 1984

British Library Cataloguing in Publication Data

Environmental implications of expanded coal utilization.
1. Coal—Combustion—Environmental aspects
I. Chadwick, M.J. II. Lindman, Nils
III. Kungliga svenska vetenskapsakademien. *Beijer Institute* IV. United Nations. *Environment Programme*
V. Akademiia nauk SSSR
363.7'387 HM221

ISBN 0-08-028734-4

Printed by A. Wheaton & Co. Ltd., Exeter, Devon

Foreword

CONCERN over the longer-term future of oil supplies is causing many countries to re-examine their energy policies for the next 20–30 years. The present doubts about whether new and renewable energy sources will become widely and readily available fast enough, the delays that several governments are experiencing with their nuclear programmes, and the difficulties in implementing conservation strategies have all focussed attention on coal and gas as fuels that can act as the bridge to a new energy future.

Many countries are now examining what degree of oil replacement they can reasonably expect from the increased use of coal by the year 2000. This is likely to involve not only increases in traditional and more advanced combustion techniques but also the use of coal for conversion to gas and liquid fuel. Two of the most important factors which may inhibit increased coal availability and use in this way are, firstly, the technical, economic and engineering problems connected with methods of rapid coal extraction and trade development. And secondly, there are the technical and scientific problems of how to handle the repercussions of expanded coal use on human health and on agricultural, forest and natural ecosystems which are essential as a resource-base for Man — impacts which are time-consuming and costly to manage well.

So far, the issues that have received the strongest recognition are the economic and engineering problems of coal production. The scientific aspects of the health and environment issues have received less attention internationally but are now becoming matters of concern on a global scale. This Volume has been produced in response to this concern.

One of the tasks assigned to the Governing Council of the United Nations Environment Programme by the General Assembly of the United Nations is "to keep under review the world environmental situation in order to ensure that emerging environmental problems of wide international significance receive appropriate and adequate consideration". The International Institute for Energy and Human Ecology (Beijer Institute) of the Royal Swedish Academy of Sciences in Stockholm is required by its charter to promote research into Man's use of energy resources and the effects of this on the human environment. The U.S.S.R. Academy of Sciences, apart from being located in a country which is one of the world's three largest producers and consumers of coal, has both centrally and through its research institutes a long-standing interest in the scientific aspects of coal utilization. Our three organizations have been pleased to work together on a Programme which has led to the production of this Volume. The nature of the Programme has enabled us to draw extensively upon eastern European sources and integrate these with western coal literature. We hope that, because of this, the Volume will form a useful addition to what has been, hitherto, a rather wide-ranging literature, not readily available in one place.

Many scientists participated in this Study either by contributing background reports or by taking part in the review of the original discussion document, or both. To all of them we would like to express our gratitude. Our appreciation also goes to M.J. Chadwick and Nils Lindman who ably brought the pieces of the Study together in its final form.

For the Beijer Institute, Royal Swedish Academy of Sciences, Stockholm:	For the United Nations Environment Programme, Nairobi, Kenya:	For the U.S.S.R. Academy of Sciences, Moscow:
Professor G.T. GOODMAN Director, Beijer Institute	Dr. E.E. EL-HINNAWI Chairman, Energy Task Force	Academician A.V. FOKIN Deputy Chief Academic Secretary, U.S.S.R. Academy of Sciences

February, 1982

Preface

THE potential environmental impacts which arise from changing the balance between the various methods of energy production and use could constitute some of the most significant agents of change in the world environmental situation. One feature of the changing balance of energy production and use is the world-wide revival of interest in coal as a future major energy alternative. As well as an increase in coal combustion using existing technologies and rapidly developing ones like fluidized bed processes, for heat production and electricity generation, attention is also being focussed on coal as a source of liquid and gaseous fuels. Whilst existing environmental problems related to the present coal fuel cycle — extraction, transportation and combustion — are likely to increase in scale, additional problems are likely to arise in relation to the commissioning and functioning of coal gasification and liquefaction plants in increasing numbers to meet the shortfall in availability of existing fuels.

Considerable attention has been paid to the environmental effects of present coal extraction and utilization activities. The land disturbance effects of mining, reclamation procedures, acid mine drainage problems, problems of aerial pollution and related risks have received much consideration. In particular, local, regional and global effects of sulphur oxides are under active investigation and the global problems of carbon dioxide emissions in relation to potential climatic change are the subject of substantial concern and research effort. Much of this information is relevant to the prediction of effects arising from the construction and operation of plants producing synthetic oil or gas. However, there is concern that when, once again, momentum gathers to develop synfuels programmes, environmental implications, including human health effects, will fail to receive adequate advance attention and research.

These concerns formed the basis of an International Workshop on the *Environmental Implications and Strategies for Expanded Coal Utilization* held in Moscow, 20–24 October, 1980. This book is not an account of the proceedings of that Workshop but the product of information tabled for discussion, altered in the light of this discussion and subsequently added to and amended as new material was accumulated.

Attempts to predict the potential environmental problems arising from coal gasification and liquefaction require an analysis of the problems and information that have emerged from the study of the traditional coal fuel cycle which culminated in the utilization of coal by combustion processes. However, because of the uneven way in which this information can predict environmental hazards due to synthetic fuel production, it has not been thought helpful to adhere rigorously to a "coal fuel cycle" arrangement in the book. The structure adopted, therefore, was as follows.

Chapter 1 outlines possible future prospects for coal. This can only be done satisfactorily in the context of predictions of the future world energy scene. Almost every "energy expert" has his own, individual, favourite scenario often based on how it is thought future trends in the world's economy will develop. However, there is acceptance by most of those concerned with energy policy that coal can play a considerably expanded role in meeting energy demands. Here we have not plumped for one prediction of the level and rate of coal expansion but sought, rather, to illustrate the interplay of industrial and economic factors that will eventually determine this.

Chapters 2 and 3 address problems common to a range of end uses of coal: combustion, gasification or liquefaction. Increased coal use, for whatever purpose, will involve increased land disturbance, increased land occupation, engender an increased volume of waste products and is likely to result in the generation of a greater volume of international trade in coal, with attendant environmental problems arising from transportation and handling.

The basic principles and technology of coal gasification and liquefaction are reviewed in Chapter 4. This is the "special" use of coal with which the book is primarily concerned. The processes involved have to be understood, as well as their potential for development and adaptation, if the environmental impacts are to be realistically anticipated.

Chapter 5 attempts to display a possible conceptual framework for the assessment of environmental and health impacts of coal utilization and conversion. It is for this purpose that the Exposure Commitment Model is introduced. However, the inadequacy of our present knowledge results in two features of subsequent chapters. Firstly, in attempts to estimate the effects of the newer coal conversion technologies, it is usually necessary to resort to information from pilot plants or information gathered in consideration of coal combustion processes. Additionally, even for coal combustion effects, the data have not been sufficiently worked to give exposure commitment estimates for single, or groups of, emissions to defined populations.

In the chapters that follow, therefore, consideration is given to categories of potential pollutants in addressing possible effects to human populations and natural systems. Chapter 6 deals with trace elements from coal combustion and conversion processes; Chapter 7 with coal derived carbon compounds (mainly the polycyclic aromatic hydrocarbons — PAHs); Chapter 8 attempts an assessment of the effects on human health of these, and other emissions, particularly on workers in the industry. However, the problems that arise are not only detected at an occupational level. There are effects on communities living close to processing plants and other human health effects distanced from processing procedures through food-chain effects, residence time effects and considerations of long-distance transport. Such pathways and mechanisms involve assessments of the effects of coal utilization and conversion on natural environmental systems. These are dealt with in Chapter 9.

The overall purpose of this book is to address the risks and hazards likely to arise from developing coal conversion technologies, at an early stage. Once the probabilities of effects can be assessed it will be possible to take preventative and remedial action in the light of policies that enterprises, governments and international agencies may wish to adopt.

M. J. Chadwick
Nils Lindman

February, 1982

Acknowledgments

THE material in this book originated in three ways: primary papers and other written material were commissioned by the Beijer Institute, Stockholm, on behalf of the three co-sponsors of the Study, for a Discussion Document prepared for an International Workshop in Moscow in October, 1980; much new information was noted for incorporation from verbal statements and from discussions during the course of this Workshop; a large amount of new reference material was obtained during the Workshop and over a period of time after it, by correspondence.

The twelve primary papers, covering the broad scope of the book, were written by Marc Ippolito (coal in the future energy scene); L.V. Motorina, V.A. Ovchinnikov and A.J. Reteyum (land disturbance and reclamation); Klaus Strzodka and Richard Steinmetz (environmental effects of brown coal mining); Harald Jüntgen (coal gasification); Stansilaw Hulisz and Maria Ihnatowicz (environmental implications of coal conversion into fuel): Karel Prasek (coal conversion processes); G.C. Butler (dose commitment concept); M. Hutton (trace elements from coal combustion and conversion processes); L.M. Shabad (influence of coal conversion products on man); R.A. Wadden and A.L. Trabert (health problems from coal conversion technologies); R.E. Munn, D. Mackay and M. Havas (effects on natural systems); S.K. Bose and A.K. Bose (environmental implications of small-scale mining operations). Some of these papers have been used in substantial measure, others only in part and M.J. Chadwick (coal transport and land reclamation) and Nils Lindman (coal gasification and liquefaction technologies) supplemented this initial material.

The second source of material employed came from notes and recordings made during the discussions in Moscow between the 70 or so active participants at the International Workshop. In places, this has given rise to substantial changes being made to the earlier material. However, by far the largest number of amendments and additions have been the result of extra written material and references assembled after the Workshop. This was the result of participants identifying areas where new information was required and ensuring that this was made available from various sources. For providing such material we are mainly indebted to B.G. Bennett and J.M. Buchanan (MARC, London); R.H. Filby (Pullman, Washington State); A. Galli (EPA, Washington D.C.); A. Grübler (IIASA, Laxenburg); G. Halbritter (Kernforschungszentrum, Karlsruhe); L.D. Hamilton (National Center for Analysis of Energy Systems, Upton); E. Novakova (C.S.S.R. Academy of Sciences, Prague); L. Strömberg (Studsvik, Nyköping); J.L. Weeks (Atomic Energy of Canada, Pinawa). Many others, too numerous to mention by name, also supplied comments and written material for which we are most grateful.

In the course of writing the final manuscript for publication the editorial writers, M.J. Chadwick and Nils Lindman received much help from Dr. J.M. Hollander and Professor G.T. Goodman of the Beijer Institute, and Dr. Essam El Hinnawi of UNEP, Nairobi. During this time Caroline Hiller acted as an editorial assistant. We wish to express our gratitude to all of them for their support and to the U.S.S.R. Commission for UNEP for their assistance with preparations for the Workshop in Moscow.

The cover illustration is reproduced by kind permission of the N.C.B. Yorkshire Regional Photographic Department, N.C.B. copyright reserved.

M.J. Chadwick
Nils Lindman

February, 1982

Contents

Abbreviations and Acronyms

(see also Units, Conversions and Equivalents)

AIChE	American Institute of Chemical Engineers
BOD	Biochemical oxygen demand
BP	Benzo α pyrene
BSR	Beavon Sulphur Removal process
C	Cholanthrene
C&EN	Chemical and Engineering News
CEC	Cation exchange capacity
CIF	Cost, insurance and freight
CMEA	Council for Mutual Economic Assistance
CNS	Central nervous system
Coalcon	A United States Energy Research and Development Agency (ERDA) hydrocarbonization process
COD	Chemical oxygen demand
COED	Char Oil Development process
CONAES	Committee on Nuclear and Alternative Energy Systems
CPE	Centrally-planned economy
C.S.S.R.	Czechoslovakian Soviet Socialist Republic
CT	Concentration × time
ct	Centimes (French)
C(t)	Concentration at time t
CTPV	Coal tar pitch volatiles
DBA	Dibenz (α,h) anthracene
DBP	Dibenzo (α,i) pyrene
DGA	Diglycolamine process
DMBA	7,12-dimethylbenz (α) anthracene
DOE	Department of the Environment (in the United Kingdom) Department of Energy (in the United States of America)
E	Period of exposure
EDS	Exxon Donor Solvent process
EEC	European Economic Community
EF	Enrichment factor
ER	Enrichment ratio
F	Flux

FDA	Food and Drug Administration (U.S.A.)
FOB	Free on board
F.R.G.	Federal Republic of Germany
F-T	Fischer-Tropsch process
GDP	Gross Domestic Product
G.D.R.	German Democratic Republic
HEW	Department of Health, Education and Welfare (U.S.A.)
HTW	High Temperature Winkler process
I	Intake
ICI	Imperial Chemical Industries
ICRP	International Commission for Radiological Protection
ID	Internal diameter
IEA	International Energy Agency
IFP	Institute Français du Pétrole
IIASA	International Institute for Applied Systems Analysis
K	Proportionality constant
LC_{50}	Lethal concentration for 50% of a test population
LNTM	Linear No-threshold Model
LPG	Liquid Propane Gas
M	Mass
MAC	Maximum allowable concentration
MAF	Moisture and ash free
MC	3-methylcholanthrene
MDEA	Methyldiethanolamine process
MEA	Monoethanolamine process
MEC	Market economy country
MeOH	Methanol
MITI	Ministry of International Trade and Industry (Japan)
MVG	Medium value gas
NCB	National Coal Board (U.K.)
NIOSH	National Institute for Occupational Safety and Health (U.S.A.)
OECD	Organisation for Economic Cooperation and Development
OPEC	Organization of Petroleum Exporting Countries
ORNL	Oak Ridge National Laboratory
P	Pressure; Transfer factor
PdC	Pas de Calais
pH	Negative logarithm of hydrogen ion concentration
POM	Polycyclic organic matter
PWR	Pressurized Water Reactor
Q	Amount of pollutant
R.S.F.S.R.	Russian Soviet Federated Socialist Republic
RSH	Mercaptans
SASOL	South African Coal, Oil and Gas Corporation
Sant-H	A coal tar oil
SCOT	Shell Claus Off-gas Treatment process
SNG	Substitute Natural Gas
SNPA-DEA	Societé Nationale des Pétroles d'Aquitaine Diethanolamine process
SRC	Solvent Refined Coal
T	Temperature
TL	Tolerance limit

U.A.R.	United Arab Republic (of Egypt)
U.K.	United Kingdom of Great Britain and Northern Ireland
UN	United Nations
UNEP	United Nations Environment Programme
UNGA	United Nations General Assembly
UNIDO	United Nations Industrial Development Organization
UNSCEAR	United Nations Scientific Committee on the Effects of Atomic Radiation
U.S.A. (U.S.)	United States of America
U.S.S.R.	Union of Soviet Socialist Republics
WEC	World Energy Conference
W-L	Wellman-Lord process
WOCOL	World Coal Study
WWT	Waste Water Treatment process (by acid gas stripping)

Units, Conversions and Equivalents

(Mainly the Système International d'Unités (SI) units have been used throughout the book except where specific use in the original work quoted makes this inappropriate; equivalents and conversion factors are therefore given)

Prefixes for SI Units

Factor	Prefix	Symbol
10^{18}	exa	E
10^{15}	peta	P
10^{12}	tera	T
10^{9}	giga	G
10^{6}	mega	M
10^{3}	kilo	k
10	deca	da
10^{-1}	deci	d
10^{-2}	centi	c
10^{-3}	milli	m
10^{-6}	micro	μ
10^{-9}	nano	n
10^{-12}	pico	p
10^{-15}	femto	f
10^{-18}	atto	a

SI Units

Physical quantity	Unit	Abbreviation	Definition (or equivalent)
length	metre	m	–
mass	kilogram	kg	–
time	second	s	–
thermodynamic temperature	kelvin	K	–
current	ampere	A	–
amount of substance	mole	mol	–
energy	joule	J	$kg\ m^2\ s^{-2}$

force	newton	N	kg m s^{-2}
power	watt	W	kg m^2 s^{-3}
pressure	pascal	Pa	kg m^{-1} s^{-2}
electrical potential difference	volt	V	J A^{-1} s^{-1}
electrical conductance	siemen	S	A V^{-1}
activity	becquerel	Bq	1 s^{-1}
absorbed dose	gray	Gy	J kg^{-1}
dose equivalent	sievert	SV	J kg^{-1} Q

(Q is the 'quality factor' which is unity for β rays, X-rays and γ rays)

Other Units and Equivalents

length	foot	f	0.3048 m
	inch	i	0.0254 m
	mile	(m)	1.609×10^3 m
area	hectare	ha	10^4 m^2
	acre	(acr)	0.4047 ha
	square inch	si	6.45×10^{-4} m^2
volume	normal cubic metre	Nm^3	m^3
	litre	l	1 m^3 $\times 10^{-3}$
	standard cubic foot	scf	2.832 m^3 $\times 10^{-2}$
mass	gram	g	1 kg $\times 10^{-3}$
	pound	lb	0.4536 kg
	tonne	t	kg $\times 10^3$
	ton (US)	(t)	0.9072 t
time	hour	h	3600 s
	year	yr	8760 h
temperature	degrees Celsius	°C	K + 273.15
	degrees Fahrenheit	°F	1.8C + 32
amount of substance	equivalent	eq	mol $valency^{-1}$
concentration	parts per million	ppm	mg kg^{-1}
	parts per million (volume basis)	ppmv	μl l^{-1}
	parts per billion	ppb	μg kg^{-1}
pressure	atmosphere	atm	1.013×10^5 N m^{-2}
	bar	bar	10^5 N m^{-2}
	pounds per square inch absolute	psia	6.9×10^3 N m^{-2}

energy	calorie	cal	4.184 J
	British thermal unit	BTU (Btu)	1055 J
	therm	–	$1.005\ J \times 10^8$
	thermie	th	0.03968 therm
	tonnes coal equivalent	tce	2.88×10^{10} J
	million tonnes coal equivalent	mtce	2.88×10^{16} J
	tonnes oil equivalent	toe	4.19×10^{10} J
	million tonnes oil equivalent	mtoe	4.19×10^{16} J
	barrel	bbl	5.74×10^{9} J
power	megawatt (electric)	MW(e)	$10^6\ J\ s^{-1}$
	gigawatt (electric)	GW(e)	$10^9\ J\ s^{-1}$
	kilowatt hour	kwh	3.6×10^6 J
	gigawatt (electric) a year	GW(e)yr	$10^9\ J\ s^{-1}\ 3.15 \times 10^7$
radioactivity	curie	Ci	3.7×10^{10} Bq
	rad	rad	0.01 Gy
	rem	rem	0.01 Sv
sound	decibel	dB	comparative unit of intensity of sound

Conversions

	kg	t	ton (US)	lb
kilogram (kg)	1	0.001	0.000984	2.20462
tonne (t)	1000	1	1.12	2204.6
ton (US)	907.19	0.9072	1	2000
pound (lb)	0.4536	4.536×10^{-4}	0.0005	1

	10^6 Btu	10^9 cal	toe	tce	GJ
tonne coal equivalent (tce)	27.337	6.88	0.6370	1	28.84
tonne oil equivalent (toe)	40.185	10.128	1	1.57	42.395

Elements and Chemical Compounds

Symbol	Element
Al	Aluminium
Ar	Argon
As	Arsenic
As_4	Arsenic (crystalline or gaseous)
As_2	Arsenic forming at high temperatures: $As_4 \rightleftharpoons 2As_2$
B	Boron
Ba	Barium
Be	Beryllium
Br	Bromine
C	Carbon
Ca	Calcium
Cd	Cadmium
Ce	Cerium
Co	Cobalt
Cr	Chromium
Cs	Cesium
Cu	Copper
Eu	Europium
F	Fluorine
Fe	Iron
Ga	Gallium
Ge	Germanium
H, H_2	Hydrogen
Hf	Hafnium
Hg	Mercury
I	Iodine
K	Potassium
Kr	Krypton
La	Lanthanum
Mn	Manganese
Mo	Molybdenum
N, N_2	Nitrogen
Na	Sodium
Ni	Nickel
O, O_2	Oxygen
P	Phosphorus
Pb	Lead
Po	Polonium
Ra	Radium
Rb	Rubidium
Ru	Ruthenium
S	Sulphur
Sb	Antimony
Sc	Scandium
Se	Selenium
Sm	Samarium

Sn	Tin
Sr	Strontium
Ta	Tantalum
Tb	Terbium
Th	Thorium
Ti	Titanium
U	Uranium
V	Vanadium
Xe	Xenon
Zn	Zinc
Zr	Zirconium
AsH_3	Arsenic trihydride
$B(OH)_3$	Orthoboric acid
CH_3SH	Methyl mercaptan
CH_3COOH	Acetic acid
CH_4	Methane
C_2H_4	Ethylene
C_2H_5SH	Ethyl mercaptan
C_2H_6	Ethane
CO	Carbon monoxide
CO_2	Carbon dioxide
COS	Carbonyl sulphide
CS	Carbon monosulphide (unstable gas)
CS_2	Carbon disulphide
$CaSO_4$	Calcium sulphate (gypsum)
$Fe(CO)_5$	Iron carbonyl
FeS_2	Iron pyrites
HCN	Hydrogen cyanide (gas)
HCO_3	Bicarbonate ion
H_2O	Water
H_2S	Hydrogen sulphide (gas)
H_2Se	Hydrogen selenide
NH_3	Ammonium gas
NO_3	Nitrate ion
NO_x	Nitrogen oxides (usually mixed gases)
$Ni(CO)_4$	Nickel carbonyl
OH	Hydroxyl ion
O_3	Ozone (gas)
PO_4	Phosphate ion
$PbCl_2$	Lead chloride
PbO	Lead oxide
PbS	Lead sulphide
SCN	Thiocyanide ion
SO_2	Sulphur dioxide (gas)
SO_x	Mixed sulphur oxides
$ZnO\text{-}Cr_2O_3$	Zinc oxide-chromium oxide mixed catalyst

1

Coal in the Future World Energy Scene

1 Introduction

THE UTILIZATION of coal as an energy source, both by means of the steam engine and in the form of coke, was a major characteristic of the first Industrial Revolution. It was on coal that nations built their industrial power in the nineteenth and early twentieth centuries. As early as 1860, Great Britain was producing over 80 million tons of pit-coal, and it is probably not an exaggeration to say that, with the control of the oceans and overseas dominions, coal made her the dominant industrial power for several decades. Until the First World

War it was the main coal-producing nations that held the world's key positions. In 1913 Great Britain, Germany, France and the U.S.A. between them produced almost 90% of the coal extracted in the whole world, and this production accounted for virtually all the world's energy supply. Throughout previous decades a 3% annual growth rate bore witness to the vigour of the industrial boom. Oil, however, was beginning to appear on the American market, and its use as a fuel was inconspicuously preparing the way for what was later to be one of the great mutations of the energy market.

Subsequently, the world entered a long period of disorder and trouble. There were two World Wars, an unprecedented energy crisis, and a massive invasion of oil into the energy market of the United States, which became the leading producer and exporter. The result of this shift to oil was a reduction in the share of coal in the world's commercial energy consumption from 61% in 1950 to 51% in 1960 and 35% in 1970. The major transition took place in the mid-1960s. By 1967 oil had overtaken coal's global market share. This pattern was not uniform throughout the world; in the U.S.S.R. rapid industrial expansion continued to be founded on coal.

After the Second World War oil production began to develop mainly outside the industrialized World – in Latin America, the Middle East, and later in Africa. The U.S.A. became, like Western Europe, an energy importer. A new balance in the world energy market came into existence based on the transfer of the developing countries' excess production to the deficient markets of the industrialized nations and by 1970 coal accounted for about 35% of the world's total commercial energy consumption (UN, 1978).

There was an exceptionally rapid and regular growth in the wealth of both the CMEA and the OECD countries. World energy requirement increased with this rapid economic progress at about 4% per year for the 1950s and 1960s, but total coal production at only 2% over this period. Solid fuels now account for no more than 30% of the total world energy supply and only 21% in Western Europe (AGIP, 1980). The leading role on the developing world energy scene was played by liquid hydrocarbons, with new discoveries being made and decreasing prices mirroring the abundance of supplies.

If this history of global primary energy substitution between 1860 and 1975 was used to project market shares into the future, coal would be credited with only 10% by the turn of the century (McDonald, 1981) but something of an energy crisis arose in the early to mid-1970s. Between 1973 and 1979 the market price of oil rose, in real terms, several-fold, producing a severe challenge to the energy balance which had prevailed for 20 years. The supply conditions of the industrial world have thus been radically changed and the price increase accentuated imbalance of the international monetary system. The situation is such that it is not now possible to envisage the uninterrupted economic growth rate of industrialized societies, predicted by a number of development models.

2 The Energy Crisis and New Economic and Demographic Prospects

The energy crisis prompts a reconsideration of the three main foundations of development models of previous years. Firstly, the basic assumptions have changed. In the future it can no longer be assumed that the industrialized countries will be the exclusive areas of growth. Secondly, the supply system is no longer ensured. Abundant, low-cost energy is a thing of the past as countries which are in possession of oil are now able to manage it as a finite, exploitable resource. Finally, the stability of the monetary system is threatened. Inflationist tensions accompany the growth in international monetary availability. Some experts have in fact set up a close parallel between soaring energy prices and monetary drifts, mutually feeding each other. The destabilization of the international monetary system did indeed begin before 1973, with the inconvertibility of the dollar dating back to 1971, but the energy crisis amplified this to a considerable extent. Under such conditions it is advisable to emphasize global balance, which is the prime condition for a stable development picture.

Several studies have been made to estimate future energy demand in the world. It should be noted that these and other energy "forecasts" rest on many different assumptions and aggregating procedures that make them only *roughly indicative* and subject to a wide range of uncertainty. The main elements affecting these are: (a) world and regional forecasts of economic development; (b) correlation of economic growth with energy consumption; (c) physical, economic, environmental and geopolitical constraints applying to energy production and consumption; (d) future prices of different sources of energy; (e) future availability of different sources of energy and the development of appropriate technologies to harness them; (f) public acceptance of the energy source and conservation measures. In addition, an inherent deficiency in all these "forecasts" is that they are currently unable to estimate non-commercial sources of energy which constitute a major resource in developing countries. Even by concentrating only on commercial sources of energy a wide range of uncertainty is encountered in assumptions pertaining to developing countries, especially in relation to the rate of substitution of non-commercial sources of energy by commercial ones.

Bearing this in mind, the selection of a particular scenario is important as the relative stability of overall balances, especially in the energy field, depends on the scenario chosen. Achieving long-term stability implies favouring development of the most impoverished countries, and in this respect a new code of development ethic is involved. Table 1.1 summarizes the relevant historical data and gives projections to the year 2000 of population, GDP and primary energy consumption. Other tables (1.2–1.5) give more details of GDP projections, energy demand, global primary energy sources and coal production and usage. The information is based mainly on the IIASA energy study (Häfele, 1981; Häfele *et al.*, 1981; McDonald, 1981) and the WOCOL study (Wilson, 1980) but other studies have also been taken into account (CONAES, 1980; EEC, 1978; WEC, 1978).

The projections presented make use of two scenarios suggested by the International Institute of Applied Systems Analysis, Energy Systems Programme Group, recently published in some detail (Häfele, 1981; Häfele *et al.*, 1981; McDonald 1981). These scenarios are defined by two basic development variables: population and GDP. In order to take into account regional differences and interdependencies, seven world regions have been defined, mainly on the basis of economic status and resource availability. They are:

- I (NA) North America (developed market economies and rich in resources).
- II (SU/EE) The U.S.S.R. and Eastern Europe (developed centrally planned economies and rich in resources).
- III (WE/JAUS) Western Europe, Australia, Israel, Japan, New Zealand and South Africa (developed market economies but poorer in resources than I and II).
- IV (LA) Latin America (developing region with market economies and many resources).
- V (AF/SEA) South and Southeast Asia, and sub-Sahara Africa excluding South Africa (developing regions with mostly market economies but with relatively few resources).
- VI (ME/NA) The Middle East and North Africa (economies in transition and with rich oil and gas resources).
- VII (C/CPA) China and other Asian countries with centrally

planned economies (developing regions with only modest resources).

It is sometimes useful to distinguish between I, II, III and IV–VII to give the three categories of developed economies and the fourth as developing nations (Table 1.1).

Population and economic growth assumptions drive the energy demand model which in turn generates secondary energy demands. This is associated with a supply model that takes account of constraints on resources and build-up rates, providing a supply mix of primary energy for each region. The IIASA model goes on to take account of investments and macroeconomic implications, but these points are only tangentially considered here.

In relation to the place of coal in these scenarios, it is necessary to emphasize one point. The traditional markets for coal have been principally for use in direct combustion to generate heat for electricity production and industrial processes, and for manufacture of coke for the iron and steel industries. In future, major new markets will develop as new technologies are perfected for direct burning of coal by efficient and relatively clean coal power cycles, and for conversion into synthetic liquid and gaseous fuels that can substitute directly for natural petroleum and gas, especially in the transport and building sectors (Hollander, 1981).

3 Coal in the Future World Energy Scene

The energy primary consumption estimates given in Table 1.1 are taken from the IIASA studies and takes into account the growth rates of GDP (Table 1.2) and energy consumption. The same tendencies can be observed, in a more pronounced way, on the basis of the world regions defined on the basis of economic status and resource availability (Table 1.5).

For an overall consumption of 18.1×10^9 tce, which constitutes the IIASA high-growth scenario estimate and corresponds to an annual growth rate of 3.2–3.3% from now till the end of the century, consumption in the western industrialized countries is assumed to be less than double by 2000, whereas that of the eastern bloc more than double and that of the developing countries to increase by a factor of three. The share of the latter increases from the current 17% to around 30% by the year 2000. The share of the western industrialized nations drops from over 60% to less than 50%. Realistic models have to take account of both a reduction in the growth of energy use,

TABLE 1.1 *Population, GDP and primary energy consumption with ranges representing low and high growth scenarios (Häfele, 1981; Häfele* et al., *1981; McDonald, 1981)*

Region	Population (10^6)		GDP/cap $1000 (1975 value)		Primary energy consumption (tce/cap)		Primary energy consumption (10^9 tce)	
	1975	2000	1975	2000	1975	2000	1975	2000
[a]I (NA)	237	284	7.0	10.7–14.4	12.25	12.7–14.8	2.85	3.6– 4.2
II (SU/EE)	363	436	2.6	5.5– 6.2	5.45	8.2– 9.2	1.98	3.6– 4.0
III (WE/JAUS)	560	680	4.3	6.5– 8.9	4.34	5.3– 6.8	2.43	3.6– 4.6
Developed countries	1160	1400	4.3	7.0– 9.2	6.26	7.7– 9.1	7.26	10.8–12.8
IV (LA)	319	575	1.1	1.6– 2.2	1.15	1.7– 2.4	0.37	1.0– 1.4
V (AF/SEA)	1422	2528	0.2	0.4– 0.5	0.25	0.5– 0.6	0.35	1.2– 1.5
VI (ME/NA)	133	247	1.4	2.6– 3.6	1.05	2.4– 3.2	0.14	0.6– 0.8
VII (C/CPA)	912	1330	0.4	0.5– 0.7	0.54	0.8– 1.1	0.5	1.1– 1.5
Developing countries	2786	4860	0.4	0.7– 0.9	0.48	0.8– 1.1	1.35	3.8– 5.4
WORLD	3946	6080	1.5	2.1– 2.8	2.18	2.4– 3.0	8.61	14.6–18.1

[a]Regional abbreviations given on pages 4-5

TABLE 1.2 *Historical and projected growth rates of GDP (percentage per year) for IIASA scenarios (Häfele, 1981)*

Region	Historical		Low scenario		High scenario	
	1950–60	1960–75	1975–85	1985–2000	1975–85	1985–2000
[a]I (NA)	3.3	3.4	3.1	2.0	4.3	3.3
II (SU/EE)	10.4	6.5	4.5	3.5	5.0	4.0
III (WE/JAUS)	5.0	5.2	3.2	2.1	4.3	3.4
IV (LA)	5.0	6.1	4.7	3.6	6.2	4.9
V (AF/SEA)	3.9	5.5	4.8	3.6	5.8	4.8
VI (ME/NA)	7.0	9.8	5.6	4.6	7.2	5.9
VII (C/CPA)	8.0	6.1	3.3	3.0	5.0	4.0
WORLD	5.0	5.0	3.6	2.7	4.7	3.8

[a]Regional abbreviations given on pages 4-5

TABLE 1.3 *Energy demand (10^9 tce) for two scenarios (Häfele, 1981; Häfele* et al., *1981; McDonald, 1981)*

Region	1975		2000 Low scenario		2000 High scenario	
	Final energy	Primary energy	Final energy	Primary energy	Final energy	Primary energy
[c]I (NA)	2.0	2.9	2.4	3.6	2.8	4.2
II (SU/EE)	1.4	2.0	2.3	3.6	2.6	4.0
III (WE/JAUS)	1.7	2.4	2.6	3.6	3.3	4.6
IV (LA)	0.3	0.4	0.8	1.0	1.1	1.4
V (AF/SEA)	0.3	0.4	0.9	1.2	1.1	1.5
VI (ME/NAF)	0.1	0.1	0.5	0.6	0.6	0.8
VII (C/CPA)	0.4	0.5	0.9	1.1	1.3	1.6
WORLD[a]	6.2	8.8[b]	10.4	14.6	12.8	18.1

[a] Regional figures may not exactly add up to total since they have been rounded.
[b] Including bunker fuel.
[c] Regional abbreviations are given on pages 4-5

TABLE 1.4 *Global primary energy (10^9 tce) by source 1975–2000 for the two IIASA scenarios (Häfele, 1981)*

Primary source	1975	2000 low scenario	2000 high scenario
Oil	3.9	5.1	6.3
Gas	1.6	2.7	3.3
Coal	**2.4**	**4.2**	**5.3**
Nuclear	0.1	1.4	1.9
Hydro	0.5	0.9	0.9
Solar and others[a]	0.21	0.3	0.3
TOTAL	8.8	14.6	18.1

[a] Biogas, geothermal, bunker fuel (1975).

especially in liquid fuels, and the replacement of oil by other fuels. Although nuclear power may penetrate some markets, in some regions, and renewable energy will expand as a source in others (Sweden, for example), there are good reasons for adopting models which attribute a reasonably optimistic role to coal (Hollander, 1981).

Whether 14.6 × 10^9 tce (low scenario) or 18.1 × 10^9 tce is assumed, the main question is can coal be developed in terms of the logistical, economic, technical and environmental problems that have to be solved? In this respect, by the late 1970s, despite all the uncertainties over oil supply and all the hopes which had been raised by various plans for 'independence', liquid hydrocarbons had retained their place in the energy balance, their weight having even increased on occasion (70 or 80% in some areas). A point has been reached where the fragility of the energy system in the industrial world keeps worsening and is heading us towards new crises whose first signs we are already experiencing.

It is clear now that nuclear energy, which according to its initial programmes would have been able to cover nearly 40% of world energy needs by the year 2000, is no longer in a position to reach these objectives, no matter how strongly these programmes may be revived. For the end of the century, the 4 million MW to be installed have been reduced to somewhere between 1 and 1.3 million MW. Converted into coal equivalents this represents 10–15% of world energy requirements, even if this percentage can be appreciably increased in Europe, especially in France.

TABLE 1.5 *Coal production and consumption (10^9 tce) in IIASA's scenarios compared to WOCOL projections by region (Häfele, 1981; Wilson, 1980)*

	Region I (North America)				Region II (U.S.S.R., Eastern Europe)			
Coal usage	1977 WOCOL	1985 IIASA[a]	2000 IIASA[a]	2000 WOCOL[b]	1975[c]	1985 IIASA	2000 IIASA	2000 WOCOL
Solid	0.14	0.14–0.15	0.2 –0.24	0.25–0.37	0.16	0.4 –0.41	0.4 –0.42	
Electricity	0.39	0.65–0.7	0.53–0.73	0.85–1.29	0.67	0.3 –0.32	0.38–0.61	
Synfuel	0	0	0.01–0.07	0.06–0.22	0	0	0	n.a.
District heat and cogeneration	–	–	–	–	n.a.	0.27–0.36	0.62–0.65	
TOTAL COAL PRODUCTION/ CONSUMPTION	0.58	0.86–0.97	0.75–1.22	1.28–2.04	0.87	1.1 –1.2	1.57–1.8	1.4
incl. Export	0.06	0.06–0.11	0 –0.16	0.15–0.25	0.08	0.09–0.11	0.18–0.12	0.1
Import	0.01	–	–	0.03	0.04	–	–	n.a.

[a] The range of figures corresponds to IIASA low and high scenarios.
[b] The range of figures corresponds to WOCOL case A (moderate increase in coal production) and Case B (big increase in coal production).
[c] WEC (1978), U.S.S.R. and Poland only.

TABLE 1.5 *continued*

Coal usage	Region III (Western Europe, Japan, Australia, South Africa, New Zealand)				Region VII (China, Centrally planned economies – Asia)			
	1977 WOCOL	1985 IIASA	2000 IIASA	2000 WOCOL	1977 WOCOL[e]	1985 IIASA	2000 IIASA	2000 WOCOL[f]
Solid	0.27	0.23–0.25	0.24–0.31	n.a.	0.31	0.38–0.43	0.48–0.62	
Electricity	0.25	0.58–0.61	0.77–0.8	0.56–0.71	0.06	0.06–0.07	0.11–0.17	
Synfuels	0	0	0.03–0.04	n.a.	0	0	0	n.a.
District heat and cogeneration	–	–	–	–	–	–	–	
TOTAL COAL PRODUCTION/ CONSUMPTION	0.52	0.8 –0.86	1.04–1.14	0.99–1.38	0.37	0.44–0.51	0.59–0.79	1.45
incl. Export	0.06	–	–	–	–	–	–	0.03
Import	0.13	0.16–0.22	0.18–0.28	? –0.31[d]	–	–	–	–

[d] WOCOL Case B: net import balance for countries in IIASA region III.
[e] People's Republic of China only.
[f] WOCOL Case B, People's Republic of China.
n.a. = not available.

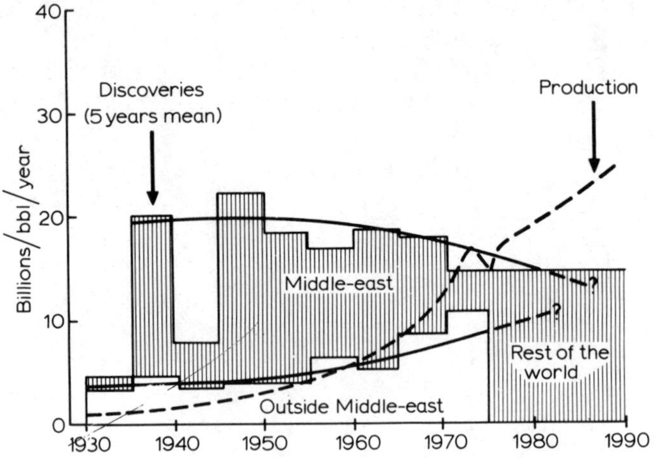

FIGURE 1.1 *Discovery rate of world oil reserves (excluding Eastern Europe and U.S.S.R.* (EXXON, 1981)).

There is, therefore, at least a transitional period to be bridged. At best, along with hydroelectric power, which will retain its relative share, renewable sources of energy may represent about 10% of the balance. Thus, there will be an "energy gap" and a real supply problem which most countries, if not the whole world, face. This must be largely met by two sources: new supplies of liquid hydrocarbons and coal. The objective must be to reduce the share of hydrocarbons in the world energy balance. If this is not achieved, the oil servitude will be enforced with attendant political risks and, within a relatively short time, geological risks as well, as is shown in Figure 1.1. The discovery curve crossed the production curve a few years ago, at least for the western world, and henceforth consumption exceeds discovery. The investment effort which has been made by the non-OPEC countries has not yet been sufficient to counter this decline in discoveries. Coal, therefore, both as a means of substituting for liquid fuel, and as a source of it, eventually has a crucial role to play over the next few decades. Increasing coal use is not just an option but a necessity. It is coal which will enable governments in the industrialized countries to recover a certain control over supply by the turn of the century. Consumption increases between 1975 and 2000, as predicted in Table 1.3, will need to be covered by up to 50% by coal (Table 1.4). The main condition

TABLE 1.6 *Breakdown of projected world energy consumption (10^9 tce), 1975–2000 (from UN, 1978; Wilson, 1980)*

	1975							2000						
	Oil	Gas	Nuclear	Hydro	Coal	New energies	Total	Oil	Gas	Nuclear	Hydro	Coal	New energies	Total
Developed countries	2319	1039	107	339	1107		4911	2350	1240	1520	600	3010	240	8960
(Ia & III)	807	234	37	130	368		1576	935	355	620	220	780	90	3000
Centrally planned economy countries (II)	676	408	6	66	1343		2499	1545	975	345	185	3690	140	6880
Including CPE Asia (VII)	93	5	–	18	514		630	905	105	85	65	1771	70	3000
Developing countries (IV–VI)	501	113	2	78	150		844	1805	485	335	515	800	220	4160
WORLD	3496	1560	115	483	2600		8254	5700	2700	2200	1300	7500	600	20 000

"New energies" refers mainly to energy derived directly from sunlight and indirectly in the form of wind, waves, tides, ocean thermal gradients or fuel from biomass; also geothermal energy.

for such a strategy to succeed is that it will need to be carried out on a large scale.

It may seem surprising that the relative share of coal in the supplies of the developing countries is the lowest and that of oil the highest (Table 1.6). But it is out of reality that this option for a high-growth scenario is proposed. It takes account of the oil potential of Asia and Latin America and also does not project a coal contribution which may be difficult to achieve in the short term. But the share of coal would have to be increased in the event that oil was found lacking.

The energy future presupposes a supply system with enough diversification and flexibility to guarantee its stability. Coal has a number of major advantages which justify recognition of its development potential (Hollander, 1981):

(a) new technologies are being developed that will allow it to be burned directly with a higher efficiency and lower environmental impacts (e.g. fluidized-bed combustion);
(b) coal can be used more efficiently and cleanly in the combined cycle approach to electrical power generation;
(c) it can be utilized for the cogeneration of electricity and steam, thereby increasing the efficiency of use;
(d) it can be developed as a source of synthetic gaseous and fluid fuel.

4 Development Prospects for the Coal Market

After more than 20 years of relative decline, especially pronounced in the western world and particularly in Europe, coal still accounts for 30% of the world's energy supply, 40% of electricity production, and 70% of the fuel for traditional thermal power stations. It is also the virtually exclusive supplier for the iron and steel industry. Coal supplies to industry and domestic households account for nearly 40% of its market.

The predicted production and consumption of coal supposes that tendencies observed since 1974 will continue to accelerate. Until 1974 energy growth was of the order of 5% per annum. Coal growth was only 2% per annum. Since then, for the first time in some 25 years, the annual development rate of coal consumption and production has surpassed that of energy supplies (3.5% as opposed to 2.7%). This has taken place in a depressed market, and would probably have been more pronounced in an expanding energy market. Predictions, in relation to recent developments, imply an annual growth rate of 2.6–3.3% for energy and an annual coal development pattern of

TABLE 1.7 *Coal production (10^6 tce): assumptions for the high-energy supply scenario (Häfele, 1981; Häfele et al., 1981; McDonald, 1981)*

Region	Base year 1975	1985	2000	2030
I (NA)	600	970	1600	2900
III (SU/EE)	870	1400	2600	3800
IV (WE/JAUS)	500	650	860	1100
V (A/SEA)	130	240	490	1400
VII (C/CPA)	350	860	1600	5800
WORLD	2450	4120	7150	13 000

2.6–4.3% (Table 1.7). This could mean more than doubling consumption and production by the end of the century and increasing the world trade in coal by a factor of five.

Coal consumption must continue developing faster than energy consumption in order to solve the supply problem on a world-wide scale. The greatest possible effort in this must be made by the leading consumers of Europe, Japan and the U.S.A. This pattern of coal growth seems to be justified by an analysis of the sectors (Wilson, 1980), summarized in Table 1.5.

4.1 Coal Consumption

4.1.1 Electricity

Until recently, electricity consumption has increased by 7% annually in western countries and this rate exceeded in Eastern Europe and U.S.S.R. centrally planned economies and the Third World. The developed countries have accounted for 90% of the increase in consumption and growth. It is this set-up which might well be drastically changed in the future as a result of the crisis.

Forecasts put the growth rate of requirements at 2.4–3.3% per year now rather than the higher figures previously given. This would mean an electricity generation of about 2000 GW(e) yr . yr^{-1} in the year 2000. However, mean projections should not conceal the likelihood of a considerable breakthrough by the Third World countries whereas the consumption rate of developed countries, both East and West, may fall off. Except for CMEA, which may prove to be autonomous, the Third World countries alone could claim almost a quarter of consumption developments for electricity, even more than a third if Asian countries with a centrally-planned economy are included. This would bring their share in supplies from 9% today up to 22% in the year 2000, or 30% if CPE Asia is included.

Installed nuclear capacities will not be as great as initially planned and thus not more than 1000 GW(e) yr . yr^{-1} can be expected from nuclear energy by the year 2000. Hydro-electric power is not expected to make a substantial contribution to electricity generation in developed countries but in developing countries electricity generation might develop to a mix of roughly one-third coal, one-third nuclear and one-third hydroelectric power by 2030 (Häfele, 1981). Any shortfall in nuclear capacity would most probably have to be met by coal. It can therefore be said that until the end of the century, and perhaps even beyond the year 2000, coal will be indispensable for roughly half of the production of electricity, a sector which itself will represent about half of the coal market.

This evolution is already under way on the international community scale. In the EEC, reservation of at least 70% of electricity supplies for coal and nuclear energy has been recommended. These figures are based on consumption hypotheses as well as on nuclear programmes (70 GW(e) in 1985, 120 GW(e) in 1990 and 300 GW(e) in 2000). Thus, in this one sector alone, 300 to 400 mtce will be required (as opposed to 150 mtce today) in order to feed capacities of 200 to 250 GW(e), and this implies a construction programme of the order of 10 000 MW(e) per year. The increase in coal consumption can already be seen in European power stations: +40% from 1975 to 1979, whereas electricity production increased by only 20%. The same phenomenon is perceptible in the U.S.A., where coal has remained the principal feed, supplying over 60% of the fuel for classical thermal power stations. Its consumption in this sector is increasing today at a rate of 10% per year (35–40% since 1975). For the year 2000, the U.S.A. is relying on coal to meet almost 60% of its electricity requirements, which, according to the government, would involve setting up over 700 power stations by the end of the century, 250 of which have already been programmed. Japan for her part will have to resort to coal increases if she is to meet consumption requirements.

The rise of coal is also inevitable in the developing countries where, except for India, it only accounts for 10% of electricity production. In some Southeast Asian countries coal power station projects planned are supposed to reach 25 000 MW(e) by 1990 as opposed to 2000 MW(e) today. This would bring coal consumption in these countries up from 3 million tonnes to 60 million tonnes over the next 10 years.

It is well known that coal plays a considerable role in the electricity supplies of CPE countries: 70–90% according to the area. Its share will remain above 70% at the turn of the century (1500–2000 mtce).

4.1.2 Iron and steel

Three main stages mark the evolution of the world iron and steel situation.

A period of prosperity from 1950 to 1975 was characterized by a rate of development in keeping with economic growth, although considerable variations in the growth rate could already be seen from one area to another. Europe was close to the world average (5%), and Japan broke all records with 14% a year. Between these two areas, the American market was already showing signs of saturation as consumption there was developing at 1.8% annually. Paradoxically, however, considering an even lower increase in production, the U.S.A. was to become one of the major import markets in the industrialized world. During this whole period, the developing countries were making a noticeable appearance on the world scene, but starting from near zero, their share remained low (10% of world consumption), and their production even lower. They therefore constituted, along with the United States, one of the great import markets.

It must be noted, however, that whereas world trade was concentrated on the export level in the EEC and Japan, it was relatively scattered as far as importers went, even beyond the U.S.A. and the less-developed countries, which represented in nearly equivalent proportions half of the world import trade. Competition between Europe and Japan was intensified, seeking for market outlets. Japan was better equipped with a more modern and competitive production tool, a more stable manpower, a more highly elaborated supply strategy, and a more active business acumen.

The second stage is characterized by the crisis and its effects, felt more acutely in Europe, but also in the U.S.A. and Japan. The crisis marks a real breaking point. A certain drift in growth rates had, of course, been noted in the preceding period, but this was natural owing to the gradual saturation of markets. However, since 1973–74, consumption and production have broken down to such an extent that none of these countries, except for Japan, can hope to recover the full use of its capacities until 1983–85. Thus it will have taken 10 years to get back to the consumption level of 1973–74. Beyond that time, uncertainties remain.

In this way we reach what might be the third stage: progress towards the year 2000. As a first approximation it seems that a scenario for future coal use can be sketched out. In the industrialized countries, the growth rate will be distinctly lower than it used to be at about 2.5% per annum. The best development possibilities seem to lie in the underdeveloped countries. According to the United Nations Industrial Development Organization (UNIDO), the requirements of these countries will evolve so quickly that they might surpass their production capacity, even though the latter is developing extensively. This in no way precludes their being competitive in certain markets for certain products, which is already the case, and these countries will nevertheless account for 25% of world consumption (35% if China is included). These figures are coherent with other energy and economic forecasts. In short, there will be a change in the balance of world energy demand with the Third World requiring a larger share. The trade balance will ultimately depend on the potential of developed areas like Europe and Japan to export their excess, which will be substantial despite their low production development, to the outlets opened up by the development of the Third World. These outlets will be mainly in Asia and Africa, alongside which the U.S.A. will remain a relatively uncertain outlet. Thus the imbalance between the developing world's consumption and production might constitute the driving force behind world trade. Outside of this hypothesis, and if specific outlets are not sought, there is hardly any future for the western iron and steel industries.

What will be the consequences of the above analysis on the coke and coking coal supply policy of the different world areas? As a first approximation, for a world steel consumption of some 1600 to 1700 million tonnes in the year 2000, and given a pig-iron/crude steel yield of 70–75%, coke requirements, which are now at a ceiling between 270 and 300 million tonnes, might reach 550–600 mtce in the year 2000, based on a coke ratio of about 450 kg $tonne^{-1}$ of pig iron. This corresponds to a coking coal supply of about 800–900 mtce as opposed to 380–400 mtce today. These tonnages include Latin America, Africa and Asia to some 25% (200 mtce instead of today's 160 mtce); the remainder would be taken mainly by the developed countries (350 as opposed to 230–240 mtce at the present time). The proportion of these supplies for international exchanges would increase from 30% or 120 mtce today to 45% or 400 mtce by the year

2000, constituting one of the foundations of world coal trade. However, when predicting future demands for coal for the steel industry it must be borne in mind that many countries are restructuring their steel processing plants away from blast furnaces to a technology that employs direct iron restoration in electric melting furnaces. This could give rise to the demand for coke being overestimated. This would mean adjusting predictions away from coke consumption in favour of increased electricity demand by the steel industry.

4.1.3 Other sectors: industry, district heating, conversion and chemical feedstocks

The stakes involved in a coal strategy cannot be adequately evaluated unless the other outlets are considered, especially domestic utilization and industry. In the developed countries coal now supplies only a small percentage of the consumption of these large sectors. Industry and domestic applications (residential and commercial heating) account for some 40% of world energy consumption, 3000–4000 mtce at the present time. But coal supplies only 30% of these requirements with 1000 mtce, 10% of this being taken up by Eastern Europe and the U.S.S.R. To substitute into this market would require a tripling of present consumption to 3000 mtce.

If the conclusion that the structure of energy consumption will not have undergone any fundamental changes by the year 2000 is accepted (WEC, 1978), it can be estimated that the industry and heating sector will call for some 8000 mtce, out of which 3000 could be supplied by coal. All of the predictions for Western Europe agree with this estimate. Out of 3000 mtce at the turn of the century, some 900–1000 would be consumed by these outlets, and with its 30% share, coal would account for about 250 mtce, compared with 50 mtce today. In the U.S.A. where industry and households will probably consume over 1500 mtce, it would seem reasonable to attribute between 500 and 600 to coal. Thus, the OECD countries will have to increase the consumption of coal for their heating and industrial energy requirements from today's 7% or 8% to nearly 30% by the year 2000.

The transformation outlet for coal utilization must also be considered. According to current projects, the price bracket for gas or oil obtained from coal corresponds closely to the present price range of the oil market: US $25–40 per barrel. Thus, it is not only to a large extent tomorrow's electricity, but also potentially tomorrow's

TABLE 1.8 *Distribution of world coal resources and reserves by country, coal quality and type of mine (Wilson, 1980)*

Region	Geological resources (10^9 tce)			Technically and economically recoverable reserves									Coal production 1977 (10^6 tce)	Reserves/ production ratio 1977 (number of years)
				Hard coal			Brown coal			Total				
	Hard coal	Brown coal	Total	Strip (%)	Under-ground (%)	Total (10^9 tce)	Strip (%)	Under-ground (%)	Total (10^9 tce)	10^9 tce	Including coking coal	Revised total (10^9 tce)		
U.S.A.	1190	1380	2570	17	83	113	49	51	64	178	18	360	623	286
U.S.S.R.	3993	867	4860	n.a.	n.a.	83	n.a.	n.a.	27	110	17	300	516	213
China	1425	13	1438	n.a.	n.a.	99	n.a.	n.a.	n.a.	99	20	300	500	198
Australia	214	48	262	22	78	18	100	–	9	27	12	75	82	329
South Africa	58	–	58	56	44	27	–	–	–	27	3	60	81	333
Poland	121	5	126	–	100	20	100	–	1	21	4	40	197	107
Canada	96	19	115	80	20	9	100	–	1	9	n.a.	10	27	333
India	56	1	57	40	60	33	100	–	–	34	3	40	100	340
U.K.	164	–	164	5	95	45	–	–	–	45	n.a.	–	123	366
West Germany	230	17	247	–	100	24	100	–	11	34	14	–	121	281
Other countries	178	50	228	n.a.	n.a.	21	n.a.	n.a.	31	52	n.a.	–	405	128
TOTAL WORLD	7725	2400	10 125	21	79	492	62	38	144	636	91	1400	2775	229

n.a. = not available

gas and oil; it is also, of course, a source of basic chemical substances.

The geographical distribution of predicted consumption by the year 2000 is well balanced in comparison to oil: approximately 1.5 × 10^9 tce in each of the main coal-producing regions (Table 1.5).

4.2 Coal Reserves and Production

The potential of coal to meet this demand is considerable (Table 1.8). Coal reserves are in order of magnitude greater than other energy reserves (WEC, 1978). Continued increase in the cost of energy since 1974 leads to projections of the exploitable potential of between 1.5 and 3.0 million mtce, under conditions of exploitation which would not be fundamentally different from those of today. It is indeed inconceivable that the cost of energy could increase ten-fold without the level of technically or economically exploitable reserves, being assessed upwards in excess of 30%. Furthermore, many experts feel (IEA, 1978) that evaluations of 600 000 mtce (WEC, 1978) are too low (see WEC, 1980). If it is true that existing geological resources amount to at least 10 million mtce (Leontief, Carter and Petri, 1977; Wilson, 1980) it can be inferred, using extremely moderate recovery rates of about 20%, that exploitable reserves are 2–2.5 million mtce. These reserves might even be multiplied, especially to the advantage of Europe, if *in situ* combustion technology can be satisfactorily developed in the next few years bringing large amounts below 1000 m within the exploitable range.

Even though the U.S.S.R. the U.S.A. and China hold between two-thirds and three-quarters of overall reserves, the amounts available in the other coal-producing countries amount to more than twice the proven amount of world petroleum resources. On this basis, the predicted consumption will only utilize up to 5–10% of the coal capital by the end of this century. This is a solid reason for believing in the feasibility of a coal strategy.

It will be noted, as an important point for the future of coal and its competitivity, that exploitable open deposits are a very considerable part of the whole: 30%, 40%, 50%, and sometimes even more, depending on the economic area. These percentages apply to the 600 000 mtce (WEC, 1978). The development of present production shows that increases are achieved more and more in open mines: more than 60% in the U.S.A. now (44% in 1970); 38% in the U.S.S.R. (compared with 27%); 50% compared with 26% in

Australia; 83% in Canada as opposed to 64% and finally 27% compared with 20% in India.

Finally, in the future world energy balance the role that developing countries may play must not be overlooked either in production or consumption. Latin America, market- and planned-economy countries in Asia, and India and Turkey will play an increasingly significant part. The influence of Eastern Europe and the U.S.S.R. will be most important and could represent half of the World's coal production and consumption.

4.3 World Trade

It is envisaged that world trade in coal will increase by a factor of five by the year 2000 (IEA, 1978; Wilson, 1980; World Bank, 1979). It is important to note that 1000 mtce (as opposed to 200 mtce today) represents 15% of predicted production (as opposed to only 8% today), a ratio which remains much more satisfactory than that which characterizes the petroleum market. The import and export potentials exist. Intra-regional exchanges (U.S.A.–Canada, EEC, CMEA) would be maintained and new exchanges within CPE countries of Asia and Latin America take place. But essential features on the import side would be Europe with approximately half and Japan and developing countries (notably in Latin America and MECs in Asia) each with a quarter. On the export side Australia, South Africa, Poland and Canada would account for 45–50%, and the U.S.A. for 35–40%. The U.S.S.R. and China, and a few developing countries (India, Indonesia, some southern African countries like Botswana, Mozambique and Swaziland, and Colombia) would account for the balance.

Everything indicates that, whereas in the past international exchanges especially involved coking coal, the situation will be different in the future. There is every likelihood that the proportions will be reversed, with steam coal occupying two-thirds of world trade, and coking coal occupying only one-third, largely for the developing countries. The volume of world coal trade doubled along with the iron and steel industry expansion between 1960 and 1974 at a much faster rate than that of coal supply and demand. Japan and Europe were responsible for 75% of the growth in imports, whereas Australia, Canada, Poland and the U.S.A. played a decisive role in the export increase. Paradoxically, this progression in exchanges seems to have been curbed since 1974. But this apparent stability conceals complex and even divergent developments. While coking coal exchanges have been generally diminishing under the effect of the iron and steel crisis, there has been a relatively strong development of steam coal exports, coming especially from South Africa but also from Australia. This tendency might well continue into the future.

In the last few years, national and international authorities, both public and private, have spoken out in favour of coal as an essential resource for safeguarding the world energy balance in the coming decades (WEC, 1978; Wilson, 1980; IEA, 1978). It is deemed desirable to set up a comprehensive coal strategy without further delay, not only to correct the energy balance but also to preserve development potential in industrialized countries, as well as allowing Third World countries to continue developing. The ideal scenario seems to be a doubling of production and consumption from now until the end of the century along with a quintupling of world coal trade.

A political will is beginning to appear in those European countries (and Japan) for which the energy and industrial stakes of a coal strategy deserve special attention for them as leading coal importers. By the end of the century these countries are likely to maintain or even to improve their status as leaders in the world coal trade. Under such conditions it is unimaginable for Europe to limit herself to spot purchases or contract buying. For this region, as well as for the other main importing areas, securing supplies under satisfactory conditions will also involve participation in all aspects of production, utilization, transport development and the acquisition of expertise in new coal-utilization technologies. Japan is probably one of the most advanced countries in this area. Ever since the crisis, MITI (MITI, 1980) has sought to transpose and adapt the main lines of supply strategy to the steam coal field with respect to the leading producing countries, especially Australia and Canada. It has become one of the world's leading iron and steel powers. It is also important to point out that large corporations are setting up and asserting their industrial and business strategies. In the past few years they have taken control of huge deposits and the ten most important among them have announced production plans for 1985–90 amounting to 600–700 mtce. This corresponds rather closely to the predictions made for coal in the expansion of world trade.

5 Economic and Strategic Aspects of Coal Development

Economic and strategic aspects are of great importance in determining whether a coal strategy will or will not pay off.

5.1 Prices and Costs

Recent estimates of production costs and prices for European steam coal are shown in Table 1.9. The estimates in this table were made in 1979. To bring this information up to date the production costs would have to be increased by at least 20%. This means that CIF prices, depending on the origin, would be in the region of U.S.$50 per ton (rather than U.S.$40) but this in no way modifies coal's competitivity, considering the evolution of prices on the oil market.

The price of coal per thermal unit today is one-third that of oil.

Steam coal is of considerable interest as a petroleum substitute but everything appears to indicate that its price is not directly related to the pressures of the general energy price situation. This is partially explainable if the rapid development of South African exports is considered. But, as the revival of the coal market becomes a reality, significant price increases can be anticipated. Beyond this situation, which some may consider to be transitory, it is important to consider whether a significant difference might remain between the price of coal and that of oil. Two basic questions arise here: What will be the cost of oil development? Will prices reflect this cost and, if so, to what extent?

Taking available reserves into account, it would only need a steam coal price increase of U.S.$10–20 per ton to create a considerable production potential. Such an increase would still leave prices considerably lower than those of oil.

But the main areas of cost rises for the future should not be overlooked. The most common objection made to the coal industry is that it is an industry based on labour which is condemned to decreasing yield and increasing costs. This is especially true of underground mines where the progress which is likely to be made in mechanization may, at least partially, compensate for cost increases. Nevertheless, the greatest hopes lie with open mines (opencast or strip mining). This is an industry which does not imply the same cost increases. If a coal strategy is to be developed, it is most likely that international channels will be supplied by either shallow or open mines.

As far as production costs go, a further constraint is the weight of transport in CIF prices. These now represent 30–50% and sometimes even 70% of cost prices in the ports of consumer areas. It is certain the use of new transport technologies could considerably reduce transport costs and thus the delivery price for large consumers.

Environmental requirements might well add further to production

TABLE 1.9 *Estimates of production costs and prices for steam coal supplied in Europe as U.S.$ per tonne, 1979 value (Charbonnages de France, 1979)*

	Production cost	Price FOB mine	Rail freight	Price FOB coast	Port charges	Sea freight	CIF price ($ t^{-1})	CIF price (ct th^{-1})	Average CIF price (ct th^{-1})	Quality characteristics: Calorific value (k cal kg^{-1})	Quality characteristics: Sulphur content (%)
U.S.A.											
Underground mines (East)	15–25	20–25	10–12	30–35	2	5	37–44	2.4–2.9	2.7	6500	1–2
Opencast mines (West)	5–15	10–20	15–20	30–35	2	7	39–44	2.8–3.2	3.0	6000	0.5
Australia											
Underground mines	10–18	15–20	5–10	20–25	2	10–13	32–40	2.1–2.6	2.4	6500	<1
Opencast mines	8–15	15–20	5–10	20–25	2	10–13	32–40	2.1–2.6	2.4	6500	<1
South Africa											
Underground mines	6–10	15	5–7	20–22	1	8–10	29–33	2.0–2.3	2.2	6200	0.5
Canada											
Opencast mines (West)	10–20	15–20	10–15	25–35	1	10–15	36–51	2.3–3.3	2.8	6600	0.5
Poland											
Underground mines				30–35	1	4	35–40	2.4–2.7	2.6	6300	<1

costs. But most studies carried out on the national and international level show that present-day energy prices should be able to cover reclamation, some desulphurization and dust control costs. According to a recent study (OECD, 1978) the cost of desulphurization and particle control would be in the order of 3–12% of the total cost of energy produced from coal. In addition, it should be pointed out that there are very large deposits of low-sulphur coal in the world.

A gap between the prices of oil and coal must be maintained in order for coal consumption to develop in industry. Technological progress in fuel–coal mixtures and fluidized bed developments give some encouragement in this direction. No matter what the drawbacks of coal may be, the delivery prices at the ports of importing countries are still competitive and so are the prices of coal in all its different uses. The cost of a coal-produced kWh is U.S. 4 cents (using coal at U.S.$50 per ton CIF). The fuel costs about half of this and the rest is attributable to exploitation costs (30%) and investment costs (70%) calculated on the basis of U.S.$650 per installed kW, including the cost of desulphurization. This cost is very close to the cost of a nuclear kWh and 55% of the cost of a heavy-fuel-oil-based kWh. Thus, it is easy to understand the movement to convert heavy fuel oil power stations to coal as the conversion costs would be recovered very quickly.

In cement-works, the transfer to coal is currently giving furnace-head economies of 45% in fuel costs. For other industrial uses (for example in boilers), the price of coal in comparison with that of heavy fuel oil and even more with that of light oil, is about 40% less, under present conditions. All these observations do not mean that prices will necessarily and continually reflect the increased costs of a sufficient remuneration margin but they do provide a hedge against continuously rising prices of oil.

5.2 *Operator Strategies*

Prices will depend to a great extent on the strategies used by the main operators: producers, consumers and intermediaries. It is by trying to find and set up a balance among the interests of these three parties that stable coal supplies at a reasonable price can be produced. Outside of this model there would be a re-emergence of dominant operators, leading to a similar situation as the current petroleum impasse. This holds true not only in production but also in transportation and distribution. Coal geopolitics lends itself better to this type of strategy than that of oil. In the producing and exporting countries involved, coal output, and more generally mining output, is an essential element of development which mobilizes a large part of the work-force. It is not as easy to close down a mine as it is to close down an oil well. The past has shown how Japan, in order to supply its iron and steel industry expansion, called upon the coal-producing industries of Canada and Australia, which had become rivals to the American coal industry – all of this under satisfactory conditions of stability. But consumer countries will still have to be present in this coal interplay in order to meet the expectations of the producing countries while drawing in the industrial and commercial counterparts which constitute the other element at stake in any coal strategy.

6 Coal Prospects

The important elements in the coal future are: the market exists; the reserves and the production potential are considerable; prices are competitive and should remain so; production, transportation and utilization techniques are all well known; port capacities can be developed to handle the increased volume of international trade; environmental problems are relatively clearly defined and capable of solution.

In a world looking for ways to re-deploy its technological capacity, coal offers a considerable potential if certain developments can be encouraged: pipelines to reduce internal transport costs; the creation of a fleet of ore ships (500 to 600 would be needed to handle the international trade on the estimated scale); the improvement and expansion of ports, in both exporting and importing countries (see Table 1.10); a boiler industry for electricity generation and industry as a whole; gasification plants for existing gas networks, the chemical industry and electricity generation; liquefaction facilities to supply petroleum substitutes and chemical feedstocks.

Looking into the future it is possible to imagine integrated sites where electricity, gas, oil and even basic chemical substances will be produced from coal. This integration offers environmental advantages. The impact of a well-developed coal strategy on world markets for the sale of goods and technology from the importing to the exporting countries is enormous. Every 1000 mtce traded generates over U.S.$50 000 million at present market prices.

From the discussion of the economic future it is concluded that coal will have to play a large part in compensating for the resource depletion and price increases of oil and gas. However, there are obstacles in the way of this transition. Liquid fuel has been the basis

TABLE 1.10 *Export and import capacities (tonnes × 10^{-6}) of world coal port terminals*

Export capacities

Countries and coal ports terminals	1978-79	2000
China	12.2	20–50
Qinhuangoao	10	
Lianyungang	2.2	
U.S.S.R. (East Coast Potential)	10	50–100
Poland (5 ports)	25	30
U.S.A.	100	350
East coast		100
Mexico Gulf		150
West coast		100
Canada	13	50–70
Australia	40	170
Hay Point	12.5	47
Gladstone	6.4	28
Newcastle	10.7	70
Sydney	2.7	4
Port Kembla	6.0	21
South Africa	25	90
Other Africa (Botswana, Swaziland, Mozambique)	0	30
Colombia	0	25–30
Indonesia		20
TOTAL	250	800–900

Import capacities

Countries and coal ports terminals	1978-79	2000
Japan	60–100	200–250
Western Europe	180	440
France	27–35	80–100
Italy	13	85
Spain	5	15
Belgium	8	20
Netherlands	5	60–100
West Germany	10	20
United Kingdom	2	20
Denmark	8	20
Sweden	–	20
Norway	1	10
Finland	5	10
Greece	0	5
Turkey		
Israel	0	5
Developing countries	10	200
Southeast Asia	3	100
Latin America	5	60
Africa		30
TOTAL	200	800–900

of much of the industrial activity during the last few decades. It is unlikely that the passage to coal as a predominant fuel can be achieved in a short time period without the appearance of certain crisis phenomena. Considerable research and development, and not a little time, is required to reconstruct the energetic basis of industry to a wider utilization of coal and to develop widely the technology of upgrading coal into fuels that can be used within the present structure of energy consumption.

The development of a coal strategy can be seen as a reply to the challenge raised by the energy crisis. If the strategy is pursued resolutely (and the political decision is the primary one from which others follow), and if the strategy is pursued to the full measure of its potential, it may make a large contribution to the solution of the present energy crisis, thus providing for future industrial and economic development.

7 References

AGIP (1980). *Eni: Energy and Hydrocarbons 1978.* AGIP, Rome.

Charbonnages de France (1979). *Communauté Etranger.* C de F, Paris.

CONAES (1980). *Energy in Transition, 1985–2010.* Report of the U.S. National Academy of Sciences Study on Nuclear and Alternative Energy Systems (CONAES). Freeman, San Francisco.

EEC (1978). Energy scenarios for the Community in the year 2000. Working document XVIII/367/78–EN. EEC, Brussels.

EXXON (1981). *World Energy Outlook.* EXXON Corporation.

Häfele, W. (1981). *Energy in a Finite World: A Global Systems Analysis.* Ballinger, Cambridge, Massachusetts.

Häfele, W., Anderer, J., McDonald, A. and Nakicenovic, N. (1981). *Energy in a Finite World: Paths to a Sustainable Future.* Ballinger, Cambridge, Massachusetts.

Hollander, J.M. (1981). Coal in the European energy future. In *The European Transition Away from Oil* (ed. by G.T. Goodman, L.A. Kristoferson and J.M. Hollander). Academic Press, London.

IEA (1978). *Steam Coal Prospects to 2000.* OECD, Paris.

Leontief, A.P., Carter, A.P. and Petri, P. (1977). *World Economy in the Future.* United Nations, New York.

McDonald, A. (1981). *Energy in a Finite World: Executive Summary.* IIASA, Laxenburg.

MITI (1980). Exploitation et importation de charbon étranger. Rapport intérimaire du Comité d'Exploitation et d'Importation du charbon étranger, Ministry of International Trade and Industry, Tokyo.

OECD (1978). *Environment and Energy Use in Urban Areas.* Organisation for Economic Cooperation and Development, Paris.

UN (1978). *World Energy Supplies: 1972–1976.* UN, New York.

UN (1978). *World Energy Supplies: 1973–1978.* Statistical Paper J22. UN, New York.

WEC (1978). *World Energy resources: 1985–2020.* IPC Science and Technology Press, Guildford.

WEC (1980). World coal trade through the year 2000 – rapid growth and political changes. *Proceedings of the 11th World Energy Conference, Vol. Ia.* IPC Science and Technology Press, Guildford.

WILSON, C. (1980). *Coal–Bridge to the Future: Report of the World Coal Study, WOCOL.* Ballinger, Cambridge, Massachusetts.

WORLD BANK (1979). *Coal Development Potential and Prospects in the Developing Countries.* World Bank, Washington DC.

2

Land Disturbance and Reclamation After Mining

1 Introduction

IRRESPECTIVE OF the actual methods of coal mining being used, mining processes unavoidably bring about land-surface disturbances, and increased mining for expanded coal supplies will inevitably lead to extended land disturbances. Such disturbed land tends to create a negative effect on the surrounding environment and reduces its economic value.

The character of the disturbance depends upon the disposition of the mined strata, and upon the methods used to mine them (Figure 2.1). Continued improvement of coal mining methods is associated

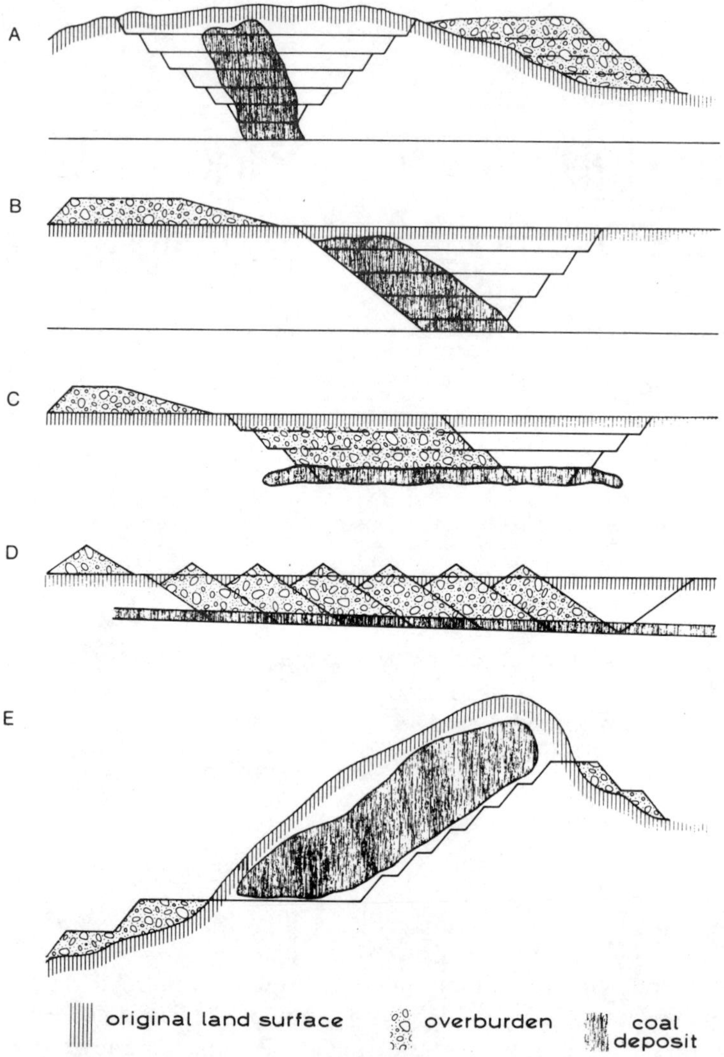

FIGURE 2.1 *Land disturbance by mining:*
A – open pit deposit deeply dipping;
B – open pit with medium dip to deposit;
C – open pit with horizontal deposit;
D – conventional area strip-mining;
E – contour strip-mining of deeply dipped deposit.

with the exploitation of deeper coal seams (below 1000 m in deep mining operations and a few hundred metres in surface- or strip-mine operations). Generally speaking, progress in mining has led to greater use of open coal extraction methods which produce larger-scale disturbances on the earth's surface. At the same time, research and development has provided greatly improved engineering and biological methods of landscape reclamation.

Although dramatic changes in landscape may sometimes occur as a result of open mining, underground coal extraction on occasion causes such gross land disturbance that reclamation runs into much greater difficulties and costs than in open mining. If geographical, social and other conditions are similar, such land characteristics as the physical and chemical properties of the rocks and soil in the rooting zone, site permeability and moisture content, predetermine the method and technology of the reclamation required. Therefore, study and classification of such properties of disturbed substrates are considered as very important.

2 Classification of Mining Land Disturbance

The following classification, of land disturbance due to coal mining, is adapted from a more general and comprehensive classification by Motorina and Ovchinnikov (1975) and Bauer and Weinitschke (1973). The classifications of Motorina and Ovchinnikov (1975) include one based on relief features and another on overburden characteristics. Although designed for the U.S.S.R., they can be adapted for other countries.

Land disturbance due to coal mining

1 Land disturbed by surface mining (strip-mining and open-pit mining).
- 1.1 Open-pit mining disturbance.
  - 1.1.1 Terraced excavations usually over 30 m deep; deposit usually dipped at >30°; overburden stored externally (Figure 2.1A).
  - 1.1.2 Terraced excavations over 10 m deep; deposit usually 8–30°; most overburden stored externally (Figure 2.1B).
  - 1.1.3 Terraced excavations usually 10–30 m deep with deposit dipped at 8°; some internal storage of overburden (Figure 2.1C).

- 1.2 Strip-mining disturbance
  - 1.2.1 Area strip-mining of horizontal deposits relatively near the surface (Figure 2.1D).
  - 1.2.2 Contour strip-mining of horizontal or steeply dipped deposits in mountainous regions (Figure 2.1E).
- 2 Land disturbed by underground (deep) mining.
  - 2.1 Land subsidence.
    - 2.1.1 Canyon subsidence caused by working medium or thick seams that dip steeply (>45°).
    - 2.1.2 Terraced subsidence resulting from the working of seams that show 27–45° dipping and where the land surface slopes.
    - 2.1.3 Bowl or cirque-type subsidence caused by the working of medium or thick seams dipping up to 27°.
    - 2.1.4 Trough-type subsidence resulting from mining low and medium thickness, horizontal or slightly inclined seams; appear like natural depressions.
  - 2.2 Surface waste deposits associated with pit heads.
    - 2.2.1 Plateau-shaped tips resulting from single- or multi-level tipping from road or rail transport.
    - 2.2.2 Crest-shaped tips resulting from cableway dumping.
    - 2.2.3 Conical tips resulting from skip or tip-wagon dumping.

Further detailed consideration of surface mining disturbance is given by Grim and Hill (1974).

3 Land Reclamation

3.1 Reclamation Objectives and Methods

The changes in the environment effected by the coal industry are so profound that often no natural recovery of disturbed landscapes can occur. Land reclamation has emerged as a method of countering the negative after-effects of extracting coal and other minerals. Land reclamation is also a comprehensive interdisciplinary subject which covers the problems of landscape redevelopment and the restoration of its productivity, ecological integrity, and economic and aesthetic value.

Disturbed lands constitute a form of environmental damage produced by the mining industry. If reclamation can be included in the production process, the mining industry could become, given a reasonable time-span, something near a "closed-cycle" industry producing no wastes. However, some overriding climatic factors may result in permanent losses, or the need for continued inputs to the system. This would be true, for example, in very high-rainfall or semi-arid regions respectively.

At the present time land reclamation has to be implemented sequentially (Motorina and Ovchinnikov, 1975). The first step involves planning, followed by research and development. The second step is site preparation for a specific use, consisting of earth-moving and shaping of the surface, water regime control and sometimes the application of productive layer, like topsoil or organic material on the surface. The third step entails biological reclamation which will allow the development of a restored site or landscape. The goal of the third step is to restore fertility and biological productivity to the disturbed land. The six main uses of reclaimed land are as follows, and depend on natural and socioeconomic factors of the locality:

(1) agriculture (arable land, orchards, meadows or grazing land);
(2) forestry (commercial or non-commercial forests);
(3) recreation (recreation zones, parks, or public open space);
(4) water use (fishing, swimming, boating and yachting);
(5) construction (sites for light industry, houses or service purposes);
(6) wildlife habitats (wilderness areas).

These uses may be closely interrelated and often form a comprehensive development in an area of extracted deposits.

3.2 *Framework of Land Reclamation Requirements*

In recent years many countries have developed and adopted laws, national programmes and specific policies for environmental protection.

The basis of most laws applicable to the mining industry and its effect on land, seeks to control land management, protect resources and regulate land reclamation and landscape restoration. Although these laws are often specific to activities of the mining industry, they constitute an integral part of national efforts to optimize living conditions for man. The costs involved in environmental improvement are borne, in the first place, by those responsible for the mining activity. There is no doubt that legal provisions of this kind contribute to rises in production costs.

The Constitution of the U.S.S.R. provides that land and its resources are owned by the Soviet society. Since the national economy is planned, the country has all prerequisites for genuinely integrated and rational use of natural resources. Article 18 of the Constitution emphasizes the importance of natural resource protection in the interests of present and future generations.

All land reclaiming activity in the country is controlled by a series of legal Acts. The fundamentals of land reclamation are decreed in *Fundamentals of Land Legislation*, the land codes of the Union Republics and other legislation. For example, Article II of the *Fundamentals of Land Legislation* makes it mandatory to reclaim the lands allocated for surface and underground mining of minerals; to bring disturbed cultivable lands and forests into a usable state; to do everything possible to restore agricultural land to a usable state in parallel with the production process, and if this is impossible to do, carry out the necessary reclamation no later than a year after production has been completed; to remove and store fertile soil layers in any work which involves soil disturbance so that it can be reused during reclamation.

An increasing concentration of disturbed land in mining areas of the U.S.S.R. has made the problem of land reclamation the most important long-term programme of the nation in the field of natural resource management. The Special Decree of the Council of Ministers of the Soviet Union of 2 June, 1976 – *On the Reclamation of Lands, Preservation and Rational Use of Fertile Soil Layers in Mining Materials and Peat, Exploration Surveys, Construction and Other Works* – provides that the reclamation of the areas where minerals have been already extracted shall be funded by the State from special allocations, and industries with active mines shall reclaim disturbed lands at their own expense, all the costs involved being covered from production costs. The research institutes of the Ministry of Agriculture of the U.S.S.R. are assigned the task of developing reclamation programmes. State control of the programme's implementation is exercised through local authorities and land management offices.

This logical sequence of laws and codes of practice in the field of land reclamation envisages a high rate of recovery of disturbed lands and during the ninth 5-year plan by 1975, the areas of reclaimed lands increased two and a half times compared with 1971. The proportion of the lands returned to agriculture increased considerably. For the Ukraine and Steppe region of the R.S.F.S.R. it reached 70–75%. This high rate of reclamation is a result of setting up special land reclamation divisions within coal production associations. Within the Ministry of Coal Industry there is a division for the protection of Nature with its own research institute. The Institute conducts comprehensive studies on problems related to reclamation of land disturbed by coal mining.

In the German Democratic Republic (G.D.R.), land reclamation was first officially endorsed in 1951 in the law on the *Restoration of Economic Value of Territory Occupied by the Mining Industry, Including Waste Storage*. At the present time various codes of practice in the G.D.R. strictly define the rights and obligations of production units whose activity causes land disturbance. Legislation determines the procedures to be followed if damage to former users occurs, including the procedure for reimbursement of losses by farmers' co-operatives when the disturbed area is greater than 20% (Motorina and Zabelina, 1968).

In Czechoslovakia (C.S.S.R.), laws concerning the use of natural resources (1957), protection of land resources (1976), forest law (1977) and building code (1976), the obligations of industrial establishments in relation to soil, landscaping, water regimes and other subjects are specified. Legislation requires long-term plans for the reclamation of lands in the large industrial areas of the country. For example, the Master Plan for the restoration of the North Czech Brown Coal Area contains detailed maps and calculations of the total land-use up to 1980. Projects provide for parks and green zones around industrial establishments. The costs involved in reclamation activity in the C.S.S.R. are borne by the mining industry.

In Poland, planned reclamation started in 1961 when the Economic Committee of the Council of Ministers made it mandatory, by a special decree for the Ministry of Mining Industry and Power, to reclaim disturbed lands. Further, a considerable effect on land reclamation activity was produced by the 1966 Act of the Council of Ministers and the 1971 Law on the *Conservation of Agricultural and Forestry Land and its Recultivation*. This legislation made land reclamation obligatory during mining operations, specified necessary procedures, and dealt with legal and financial considerations. Reclamation work is carried out by the coal producers as a cost to them.

Land reclamation is legislated for in Bulgaria, Hungary, Romania and Jugoslavia. Although there are differences in organization and management, a number of common principles can be discerned:

(1) land reclamation activity is included in national plans and constitutes a part of natural resource management;

(2) objectives and methods of reclamation must be determined at the design stage of the mine; mining plans must incorporate land reclamation as an integral part of production processes;
(3) before reclamation plans are drawn up a comprehensive study is carried out.

In the United States of America, the *National Environmental Policy Act* of 1969 covers a wide area of national policy in relation to environmental problems. The Act requires the preparation of a document to be used by the Federal authorities in their decision process for all major Federal actions that significantly affect the environment. The Federal authorities must take into account:

(1) the environmental impact of the proposed action;
(2) any adverse environmental effects which cannot be avoided should the proposal be implemented;
(3) alternatives to the proposed action;
(4) the relationship between local short-term uses of Man's environment and the maintenance and enhancement of long-term productivity;
(5) any irreversible and irretrievable commitments of resources which would be involved in the proposed action should it be implemented.

Such a document is required wherever Federal action is involved in the mining process, either direct or indirect (for example, the Federal licensing of a related activity).

Until 1977, individual States in the U.S.A. exercised primary jurisdiction over the restoration of mined lands. State legislation usually authorized a state agency to issue permits to mines on receipt of details on the mine site, mining and reclamation plans, and a bond or other security to cover the restoration.

Surface and mineral rights are owned separately and diversified ownership among private individuals, the Federal government and the State governments exists. Until 1976 it had been the policy of the Federal government to allow the individual States to act for it where Federally owned coal was being mined, but this was changed by the 1977 Federal *Surface Mining Control and Reclamation Act.* This created overall regulations for coal mine reclamation and established minimum reclamation performance standards. The standards include restoration to the approximate original contour, segregation and replacement of topsoil, establishment of vegetation comparable to pre-mining conditions, and protection from adverse hydrological effects. Operators must take responsibility for successful revegetation for 5 years after seeding (for 10 years where mean annual precipitation is less than 66 cm).

Thus, since 1977, reclamation of surface mined land for coal is initially the responsibility of the Office of Surface Mining within the U.S. Department of the Interior, but individual States are allowed to take this over where they develop State programmes that comply with the procedures in the Surface Mining Act and enforce permanent standards at least as stringent as the Federal ones. If State programmes are not developed they have no regulatory authority over the reclamation of coal mined land. The Surface Mining Act placed a levy per ton on all coal produced in the U.S.A. This fund can be used to finance the restoration of areas that were not reclaimed in the past and other action needed to alleviate off-site impacts of the unreclaimed areas.

From 1981 a new consideration has been given to the Surface Mining Control and Reclamation Act. Without amending the Act it is intended that the rules should be streamlined to reduce Federal involvement in the Act's enforcement and increase the discretion of State authorities in interpreting the Law according to regional conditions. It will probably result in a reduction in site inspection by the Federal agencies and remove certain specific requirements such as those relating to the grading of coal haulage roads and the method of treating mining water runoff. Regulations relating to the planting density of forest trees on reclaimed forest land may also be relaxed in the light of local conditions.

In the Federal Republic of Germany, considerable pre-planning of reclamation has occurred as coal mining has impinged on areas of intensive agriculture. As early as 1920, a law established an association to deal with regional planning in the Ruhr. The *Regional Planning Act* of 1962 made this body responsible for a development plan which includes refuse disposal planning and land utilization policy. Open space and amenity areas must be secured, often by the renewal of derelict areas. The location of tips is determined, suggestions for combining waste materials are made and landscaping plans and planting schemes approved. This provides integrated planning control for land restoration.

In the Rhenish brown coal district of the Federal Republic of Germany, a soft brown coal has been mined by surface methods since the 1950s. An integrated plan defines the limits of mining, the location of industry, agriculture and forestry and the reclamation schedule. This operational plan is approved by the *Land* parliament.

In the United Kingdom local government has had powers to acquire derelict land for restoration since 1944 (*Town and Country Planning Act*) and in 1951 the *Mineral Workings Act* provided for a fund for financing the restoration of land made derelict by surface mining by stripping. Various acts made provision to meet the high cost of land restoration by making grants available. These have ranged from 50% to the total cost in special development, development, intermediate and derelict land clearance areas. Since 1974, as a result of an amendment to the *Town and Country Planning General Development Order 1973*, a county Planning Authority may require the National Coal Board to submit a scheme aimed at ensuring that waste material from deep mines is tipped in such a way as to facilitate subsequent landscaping and restoration to improve visual amenity. Strip-mining land restoration has been developed extremely successfully by the Opencast Executive of the National Coal Board.

In the People's Republic of China an Environmental Protection Law is under trial implementation. An Environmental Protection Office implements and supervises the carrying out of the national guidelines, policies, laws and acts relating to environmental protection. Land protection figures prominently in many of the articles of the law.

3.3 Developments in Mining Technology and Land Reclamation

During the last 15–20 years considerable progress has been made in reclamation technology geared to the biological improvement of disturbed land and the creation of optimally organized landscapes. This generally involves:

(1) careful separation and placement of different categories of mining waste;
(2) earth movement and landscaping to meet fully the requirements of potential users;
(3) careful analysis and amendment of waste materials to ensure good plant establishment and growth;
(4) attention to drainage and other hydrological factors.

International experience shows that the site engineering stage of reclamation is the most costly part of the operation. If, however, engineering works are included in the production process and planned in the course of mining operations they may become economically effective and efficient.

3.3.1 Strip-mining techniques

Experts in Czechoslovakia believe that the opening pioneer cuts in surface mining operations can be located in such a way that, at the final stage of mining, overburden and soil material is strategically placed for the final stages of reclamation work. This approach makes it possible to provide the best soil at the surface of reclaimed land when levelling the area (Stys, 1966). When large horizontal coal deposits are mined in the U.S.S.R., G.D.R. or F.R.G. by surface mining, open pits are arranged in such a way that when each new pit is opened spoil is disposed of in the old mined-out pit. This technique provides for maximal filling of old pits and a reduction in the areas needed for external waste tips (Koliadny and Ovchinnikov, 1976). A large number of operational improvements have occurred in recent years in the U.S.A. in the surface mining of coal (Grim and Hill, 1974).

3.3.2 Improvements in site engineering technology

The degree of site engineering work required is largely determined by the desired after-use of the site. Generally, land that is to be reclaimed to arable use requires the greatest degree of site engineering.

Requirements for site engineering can most easily be met if waste and waste storage transport systems are planned in advance. Stripping that creates a "hill and dale" topography with up to 15–20 m difference in elevation requires a great volume of earthmoving to be completed (up to 20 000 m^3 ha^{-1}) if the surface is to be rendered suitable for ploughing. A reduction of levelling work can be achieved if the width of stripping face is reduced, or if waste storage is formed by fan techniques. This method of pile formation has been introduced at a number of open pits of the U.S.S.R. (Estonian shale, Cheremchovo coal) and is also used extensively in the U.S.A. (Ovchinnikov, 1967; Kaar and Lainoja, 1971; Grim and Hill, 1974).

The greatest experience in reclamation of crest-shaped waste storages has been accumulated in the U.S.A. because waste is stored internally (Figure 2.1D) without intermediate transport. The geological configurations of the coal deposits favour such methods of mining.

The technology and processes which are used in reclamation are dependent on the types of stripping and dumping machinery that are used and the way in which they are deployed. Many experts believe that draglines are the most efficient means of levelling waste surfaces. Caterpillar and wheeled dozers have also become widely accepted means of levelling surfaces in connection with both surface and deep mining coal.

Wheeled-bucket loaders of 8 m^3 or more capacity and wheeled scrapers are used on a large scale. The efficiency of scrapers in reclamation works is determined by the distance it is required to move the overburden. If the distance of earth-moving is significant scrapers can be efficient.

Reclamation work in the U.S.A. often leaves surfaces incompletely levelled and it is claimed that this provides some biological advantages in reclamation (Riley, 1973). Surface levelling can be divided into three categories: complete, terrace-forming, and incomplete (peaks having a crest width of 2.5–5 m at the top). Complete levelling results in an even horizontal surface and is rarely undertaken in the U.S.A. The most widespread technique is incomplete and terrace-forming levelling. This is much less costly, while the surface suits quite well the establishment of grazing lands and forests. In the climates with low precipitation (the Southwestern States) the surfaces of reclaimed waste storages are levelled to form closed drainless depressions which are used as reservoirs for gravity irrigation.

3.3.3 Improvements in surface soil treatment

One of the main preconditions of successful biological reclamation is a substrate for plant growth that is physically, chemically and biologically fertile. If stripping and dumping of the overburden in strip-mining operations is planned beforehand, coarse materials can be buried and overlain by fertile substrates. During stripping, toxic wastes if identified, can be selectively stripped and channelled to the bottom of storage dumps.

In the G.D.R., at a number of open pits of the Niederlausitz brown coal basin, selective topsoil stripping is ensured by using excavators and transport bridges.

In the Rhenish brown coal basin thick loess deposits are found in the northern part of the basin but in the southern part no appropriate material is present. A solution to this problem is found by transporting the loess by railroad for over 40 km. After delivery the loess is discharged hydraulically to levelled land (the "loess flush system").

The loess is applied in layers. First a 20–30 cm layer is applied and is partially mixed with the waste. After this layer has dried off it acts as an impermeable stratum and prevents water from subsequent loess applications from seeping into the pile. The water is collected and recycled. The plots are drained after the last application of loess and cultivations can take place within 1½–6 months. In addition, trees of high economic value have been planted successfully on forest gravel – a mixture of loess and sand, gravel and crushed rock (Kausch, 1979).

3.3.4 Land-use efficiency improvements

"Land ratio" in surface mining of coal is an important consideration. It is defined as the area of land taken up or disturbed to produce 1 million tons of coal. Horizontal deposits usually result in a lower land ratio than inclined deposits because the latter usually involve external storage of the waste (compare Figure 2.1B and D). For this reason studies have been made to suggest designs for internal waste storage where steep and inclined seams are mined (Tomakov, 1975). If, however, external waste storage has to be deployed on fertile land, the most effective method of waste disposal is often in single large sectional banks, rather than as high multibank stores. This method results in somewhat higher land ratios, but reclaimed surfaces can be brought into agricultural production at a faster rate as single-bank storage gives a greater ratio of horizontal to sloped surface, meeting agricultural requirements better.

3.3.5 Improvement of underground mining waste reclamation

Most underground coal mines no longer tip colliery spoil to form conical tips. The most progressive form of tipping involves the establishment of a common dumping site for a group of mines utilizing motor transport, and producing plateau-shaped or contoured tips. Such a procedure is now common in many coal-producing countries in Europe and North America.

Especially interesting is a method of forming large multibank waste tips (of a "step-pyramid" type) in Ruhr. Spoil is hauled by dump trucks to the tip site. The waste is formed from the periphery to the centre of the dumping area, thus making reclamation of the perimeter possible right from the start of dumping. A pioneer retaining bank 8–10 m high is formed at the outset, this height being maintained for the first stage of tipping. The outer slope forms part of the final tip, unchanged. This waste platform is then used for the next stage of tipping and a second similar bank constructed. A distance of 4 m between the upper edge of the first bank and the bottom edge of the second is left as an access roadway. This pattern is followed until the tip reaches the desired height. Completed tips have four to six banks (terraces) with 1 : 2 or 1 : 3 outer slopes. The outer slopes are composed of fine refuse and these are seeded or planted with trees. By the time waste tipping is completed the outer slopes are fully reclaimed. Tree growth softens the outline of the tip but recently more consideration has been given to landscaping, and areas that are to receive deep mine spoil are now often prepared to receive it by first scraping off the fertile topsoil. The area is then used as a dumping site, the volume of tip being calculated to match the surface area to be covered by soil. Some areas that suffer from subsidence can make use of this so that as soon as the coal is extracted the roof will cave in and the tip becomes flush to the surface of the earth.

Some systems have been developed which utilize mined-out underground voids for colliery spoil stowage. This reduces land subsidence to a minimum and removes the need to occupy land with spoil tips. Experiments have been conducted to find the best method of 'back stowage'. No universal approach, however, has yet been found. Moreover, if thick coal seams are mined, the back stowage material available may be insufficient in quantity and special quarries may have to be opened up to produce it. This situation exists in the Upper-Silesia Coalfield in Poland. Here 90% of the coal produced underground is mined using a system of back stowage. Land subsidence is negligible. The stowage material, however, has to be produced separately from sand open pits of 40 million m^3 capacity. Reclamation of this is often achieved by filling them with colliery spoil!

The reclamation of disturbed land is not only a problem – it is a challenge as new land-use patterns may be developed and the original productivity improved. If "land capability analogues" and "waste material analogues" can be recognized, from region to region and country to country, a system can be made available to optimize land restoration practice. However, it must be admitted that some very real problems still remain. Arid or semi-arid regions present restoration problems that are only just beginning to be squarely faced. In developing countries such problems may be particularly acute. Even more conventional reclamation situations may not attract the expertise necessary for their solution. There is a real need for restoration expertise and experience to be matched with the conditions likely to prevail in potentially new centres of coal mining activity.

3.4 Progress in Land Reclamation

3.4.1 General considerations

All reclamation will depend on such characteristics as the nature of the land disturbance, the climatic and environmental conditions prevailing, economic considerations and social requirements. For example, in regions where fertile soils occur near densely populated areas, normal requirements for restoration will be to agriculture. Less fertile regions or industrial areas that are in need of measures against air pollution will usually be restored by the establishment of forest areas. Some reclamation may make use of flooded pits to construct reservoirs, water recreation and amenity areas.

It is held by many reclamation scientists that the selection of objectives can be made more rational and cost-effective if a classification of disturbed land types is produced showing their compatibility with reclamation objectives. Classification by spoil type, relief, rock composition and characteristic colonizing vegetation of broader application have had to be developed. In Poland, six categories of disturbed lands have been identified on the basis of relief, soil, and water regime characteristics.

In the U.S.A. a coding system has been developed for disturbed lands based on aerial photography. In the U.S.S.R. a classification of disturbed lands has been developed based on landscape and technological features.

Although classifications based on straightforward mineralogical composition and the chemical and physical properties of wastes are insufficient alone to identify broad reclamation potential, they can be nonetheless extremely important. This is particularly true of coal waste materials in relation to iron pyrite (FeS_2) content and form, once the waste material is exposed to water and the dissolved oxygen it carries. Pyritic oxidation is responsible for toxic substrates due to low pH values (as low as 1.2–1.4) and subsequent phytotoxic levels of aluminium, and some other heavy metals. Nitrogen, phosphate and even potassium deficiency (Caruccio, 1975; Chadwick, 1973) can also occur. Acid mine drainage water causes pollution of aquatic ecosystems (Glover, 1975). Methods of dealing with these problems have been intensively studied.

3.4.2 Land disturbed by surface mining

Reclamation of land disturbed by mining for coal is now firmly established in most developed countries, on the scientific analysis of substrates, experimental determinations of amendment requirements and detailed planting and sowing regimes. As a result, land is restored to a wide range of after-uses. Scientifically and technically it is now possible to deal with a wide range of reclamation situations in a successful way (Motorina, 1975). This is not to say, however, that no scientific or technical problems remain, or that the application of reclamation knowledge always results in satisfactory land restoration. One of the largest environmental disturbances resulting from the surface mining of coal in Europe is to be found in the F.R.G., in the Rhenish brown coal district. Resources amount to 55×10^9 tonnes. A 175 km^2 area has been mined in an area with a population of over 1 million and a population density of 413 inhabitants km^{-2}. About 116 km^2 have been reclaimed, and 23 700 inhabitants resettled in 55 resettlement villages (Kausch, 1979). The methods of reclamation have been described in a previous section (3.3.3).

In the G.D.R. brown coal is mined in two main areas: Halle-Leipzig and Niederlausitz. This produces about 260×10^6 tons per annum. 88 900 ha have so far been mined and over half of this area restored. Reclamation is complicated in some areas by the existence of a thick stratum of sulphide-containing rock. Where fertile loess exists, this is extracted to a depth of 1 m and stored separately from the rest of the over-burden. In the Niederlausitz area infertile sands and gravels overlie clays and sands with a high sulphur content (4–6%). These clays and sands are often replaced in reclamation as the surface layer, but being acid (pH 2.8–3.5) soil improvement measures must be taken. Either lime or other neutralizing materials must be added. In the "Domsdorf" method (Knabe, 1959, 1973) applications of hygroscopic calcareous brown coal ash, from nearby power plants, are combined with fertilization and deep soil cultivation. Various modifications of the method have been worked out (Illner and Raasch, 1966). Blast furnace slag and the residue of waste-water treatment from coking plants ("Koyne" method) have also been used as ameliorants. On agricultural areas good yields of lucerne (*Medicago sativa*), rye (*Secale cereale*) and sunflower (*Helianthus annuus*) have been obtained. Poorer soils are devoted to forestry, planted initially with buckthorn (*Hippophaë* spp.), alder (*Alnus* spp.) and black locust (*Robinia pseudoacacia*) as pioneer species, followed by poplar (*Populus* spp.), pines (*Pinus* spp.), birch (*Betula* spp.) and oak (*Quercus* spp.).

Reclamation in the C.S.S.R., following the surface mining of coal in northern Czechoslovakia and the Sokolovsky region, concentrated for some years on improving the air quality conditions of industrial regions by afforestation. Red oak (*Quercus borealis*), European larch (*Larix decidua*), ash (*Fraxinus excelsior*) and field maple (*Acer campestris*), after European alder (*Alnus glutinosa*), grey alder (*A. incana*) and sea buckthorn (*Hippophaë rhamnoides*) have been used as "soil improvers". In the Sokolovsky brown coal basin apples, pears, plums and cherries, as well as soft fruit, are grown on reclaimed land.

Recently, reclamation to agriculture has been favoured and research has centred on the ways in which topsoil applications can be reduced in depth from 50 to 20 cm. This has been achieved by fertilizing and growing a green manure grass-legume mixture for the first 3 or 4 years. Subsequently good yields of wheat (*Triticum aestivum*) and barley (*Hordeum vulgare*) are gained.

Reclamation of brown coal mined areas in other Eastern European countries (Hungary, Romania, Bulgaria and Poland) has also successfully produced land suitable for a range of agricultural crops and forestry.

Strip mining for coal is practised in the U.K., where the ratio of overburden to coal (by volume) is 20 : 1 for bituminous coal and 30 : 1 for anthracite. The stripping operations in the U.K. are now planned not only to reclaim the areas currently being exploited, but also to incorporate spoil material from old deep mines. Care is taken to remove topsoil and layers of overburden, and store these separately. Care is also taken in replacement. After soil replacement cultivation is needed to restore soil structure; often high nitrogen dressings are required to restore losses from the soil dump by denitrification. Restored sites in the U.K. have generally given good yields of arable crops and satisfactory tree growth (Bradshaw and Chadwick, 1980).

Over 30% of coal in the U.S.S.R. is produced by surface mining. It is envisaged that this will quickly rise to 40–45%. As conditions of climate, geology and soil type differ enormously over the U.S.S.R. specific approaches to reclamation have to be adapted from region to region.

In the Ukrainian S.S.R., surface mining disturbs fertile chernozem soils that are used extensively for agriculture. Either the upper layers of the soil (the black A_1 horizon) are replaced on the surface to form the growth substrate, or in some areas, it has been found that the lower loess deposits are more fertile and these are held back for surface spreading. Usually legumes are sown first and where these are used as green manures high yields of winter wheat (*Triticum aestivum*) and other crops are obtained (Novikova and Savitch, 1980). Considerable emphasis is given to crop rotations.

Two areas of investigation have been pursued vigorously in recent years: the screening of toxic sulphide-containing rocks with impermeable or carbonate rocks before topsoil application, and the necessary depth of topsoil under different conditions.

Restoration of land to forest and screens of trees is undertaken. In the Ukraine 40 species of woody plants have been established on coal spoil, including elm (*Ulmus* spp.), black locust (*Robinia pseudoacacia*), lime (*Tilia* spp.), apple (*Malus* spp.), and species of pine (*Pinus*), maple (*Acer*), poplar (*Populus*) and birch (*Betula*).

Consideration has also been given to the reclamation of power plant ash (fly ash) and settler and sludge pond material. Generally these lack major plant nutrients (nitrogen, phosphorus and potassium) and are physically poor. Layers of soil, peat or organic wastes have been used, in conjunction with fertilizers to improve the chemical and physical condition. Arable and pasture crops have been grown and tree species established.

In the U.S.A. area and contour strip-mining are widely practised for coal gaining. In the areas where deposits are shallow, or there are seams of considerable width, there is less difficulty with reclamation to a satisfactory standard than where moisture is lacking, contour mining is required or high iron pyrite contents are encountered. It is estimated that 80% of the land disturbed can be reclaimed, 25% of it to agriculture. A specific feature of the reclamation activity in the U.S.A. is that the surface of the disturbed land may only be partially levelled to enhance water availability and, in some cases, reduce toxicity by the creation of a 'ravine-ridge' surface topography (Riley, 1973). Negative effects of complete levelling have been demonstrated. Considerable expertise has been built up on the technological and biological aspects of reclamation and many good summaries of all aspects of this work exist (Hutnik and Davis, 1973; Grim and Hill, 1974; Schaller and Sutton, 1978; Bradshaw and Chadwick, 1980). In particular the mechanization of reclamation has been developed and hydraulic sprinklers, hydraumatic seeding and the aerial application of fertilizer and seed have been extensively applied.

3.4.3 Land disturbed by underground mining

Although in many countries the percentage of coal mined by surface mining methods is increasing, current deep mine waste production still produces considerable environmental degradation to add to the many unreclaimed spoil tips produced in previous decades. Generally, these tips are to be found in densely populated regions and the possibility of their arrival with any major coal mining development is one of the major deterrents to public acceptance of plans for coal mining in an area.

Colliery spoil heaps can be satisfactorily restored to blend into the landscape and to be biologically productive, but in almost all countries one or several factors may prevent this easily being achieved. Iron pyrite, a primary mineral associated with colliery spoil, is responsible for rendering the medium acid. Usually acidity is produced as this is oxidized on the tip and there is seldom sufficient natural carbonate mineral, or other buffering components, present to neutralize the acidity. Spoil pH values may fall below 3. The complicated relationship between pyritic content and acid generation has been the subject of many studies and these are well summarized by Caruccio (1975). The result of pyritic oxidation on coal tips is to produce surface layers of low pH and this in turn has a detrimental effect on spoil components that renders the spoil even less suitable for satisfactory plant growth. Low pH reduces the spoil cation exchange capacity so that beneficial ions are more easily lost by leaching; clay minerals become degraded and aluminium in solution reaches phytotoxic levels; the mineral ankerite is degraded to release high levels of available manganese; clay mineral distortion may lead to the release and loss of potassium. In addition, low pH values (accompanied by high iron and aluminium concentrations) cause phosphate fixation and reduce nitrogen availability.

In some regions, as in the Silesian Coalfield in Poland, spoil may show salinity problems due to a high content of water-soluble salts originating from ground water. Salinity can also develop from weathering reactions involving carbonate minerals in the spoil, calcium, magnesium and sodium accumulating, mostly as sulphate and chloride. In humid climates these salts are relatively easily leached from spoil material and salinity problems are fairly short-lived. This is not true for more arid climates.

Spoil materials from deep-mined coal are highly variable from region to region, tip to tip, place to place on a single tip and from time to time (Chadwick, 1974) and this must be taken into account in formulating reclamation procedures.

After earth-moving and landscaping work is completed, correction of spoil reaction is usually carried out by addition of lime – even amendments in excess of 50 tonnes ha^{-1} may be required but care needs to be taken in order not to induce temporary phosphate unavailability due to high pH values. Most colliery spoils will require additions of nitrogen and phosphorus fertilizer, in amounts that can be determined by analysis, and then suitable plant species and cultivars must be selected. Legumes are often employed in order to contribute fixed nitrogen, although under-planting trees with these may cause a reduction in tree survival and growth. The various planting, seeding and spoil-amendment techniques suitable for colliery spoil in different parts of the world have been the subject of much research and a summary is provided by Zaitzev, Motorina and Danyko (1977) and Bradshaw and Chadwick (1980). In particular, it is now realized how important careful attention to maintenance procedures can be, and the part site monitoring must play in this.

4 The Economics of Reclamation

As land is taken out of circulation by mining processes and the environment degraded by the wastes of industry, sectors of the economy naturally suffer. The efficiency of investments into the mining industry must be evaluated taking into account the real costs involved. The costs of environmental protection measures have to be internalized.

In all countries the cost of the earth-moving stage of reclamation makes up 70–90% of the total costs. This is a high cost – low risk activity whereas the cultivation, sowing and planting part of reclamation is a low cost – high risk activity. Costs vary enormously and it is not possible to make sensible comparisons from country to country.

It is difficult to separate the economic effects of reclamation from widespread social effects. Some of the assessments published suggest that sociological factors far outweigh economic effects. A correct assessment of the importance of reclamation will only be possible when socioeconomic aspects are given their proper emphasis.

Finally, the economics of the total coal production and utilization cycle need to be analysed from the standpoint of energy consumption. Such a technique would evaluate the energy balance of the total coal cycle, including reclamation.

5 References

BAUER, L. and WEINITSCHKE, H. (1973). *Landschaftsphlege und Naturschutz.* Jena.

BRADSHAW, A.D. and CHADWICK, M.J. (1980). *The Restoration of Land.* Blackwell, Oxford.

CARRUCCIO, F.T. (1975). Estimating the acid potential of coal mine refuse. *The Ecology of Resource Degradation and Renewal* (ed. by M.J. Chadwick and G.T. Goodman). Blackwell, Oxford.

CHADWICK, M.J. (1973). Methods of assessment of acid colliery spoil as a medium for plant growth. *Ecology and Reclamation of Devastated Land*, vol. I (ed. by R.J. Hutnik and G. Davis). Gordon & Breach, New York.

CHADWICK, M.J. (1974). Variation in the chemical characteristics of colliery spoil and its relation to vegetational establishment. In *Grüne Halden in Ruhrgebeit*, pp. 22–38. SVR, Essen.

GLOVER, H.G. (1975). Acid and ferruginous mine drainages. *The Ecology of Resource Degradation and Renewal* (ed. by M.J. Chadwick and G.T. Goodman). Blackwell, Oxford.

GRIM, E.C. and HILL, R.D. (1974). *Environmental Protection in Surface Mining of Coal.* EPA-670/2-74-O13. U.S. Environmental Protection Agency, Cincinnati, Ohio.

HUTNIK, R.J. and DAVIS, G. (1973). *Ecology and Reclamation of Devastated Land*, vols. I and II. Gordon & Breach, New York.

ILLNER, K. and RAASCH, H. (1966). Zur Bestimmung des Kalkbedarfs für die Melioration von Schwefelhaltigen Tertiärkippen. *Zeitschrift fur Landeskultur*, **7**, 285–90.

KAAR, E. and LAINOJA, L. (1971). *Polevkivikarjaaride Rekultiveerimine.* Tallinn.

KAUSCH, P. (1979). Brown coal opencast mining and landscaping in Rhineland. *Energy Research*, **3**, 275–88.

KNABE, W. (1959). Vorlaufige Richtlinien für die Bepflanzung und Melioration forstlicher Kippenstandorte in der Niederlausitz. *Forst. und Jagd.*, **9**, 229–34.

KNABE, W. (1973). Development and application of the "Domsdorf ameliorative treatment" on toxic spoil banks of lignite opencast mines in Germany. *Ecology and Reclamation of Devastated Land*, vol. II (ed. by R.J. Hutnik and G. Davis). Gordon & Breach, New York.

KOLIADNY, M.F. and OVCHINNIKOV. V.A. (1976). *Rekultivatsia Narusiennich Ziemli v FRG.* Moscow (Russian).

MOTORINA, L.V. (1975). *Experience of Reclamation of Land Disturbed by Mining in the USSR and Abroad.* Moscow.

MOTORINA, L.V. and OVCHINNIKOV, V.A. (1975). *Industry and the Recultivation of Land.* M'isl, Moscow (Russian).

MOTORINA, L.V. and ZABELINA, N.M. (1968). *The Recultivation of Land Damaged by the Mining Industry – A Review of the Literature.* All Union Scientific and Technical Institute for Agriculture, Moscow (Russian).

NOVIKOVA, N.A. and SAVITCH, A.I. (1980). *Vosstavnovlienie Potivennovo Plodorodia na Otvaliach Otkrutuch Razrabotok pri Biologitieskoy Rekultivatsii.* Moscow.

OVCHINNIKOV, V.A. (1967). The interrelationship between natural and technical mineralogical (mining) factors in the recultivation of land damaged by opencast mining operations. In *The Restoration of Land Following Industrial Operations.* M'isl, Moscow (Russian).

RILEY, C.V. (1973). Chemical alterations of strip-mine spoil by furrow grading – revegetation success. *Ecology and Reclamation of Devastated Land*, vol. II (ed. by R.J. Hutnik and G. Davies). Gordon & Breach, New York.

SCHALLER, F.W. and SUTTON, P. (1978). *Reclamation of Drastically Disturbed Lands.* American Society of Agronomy, Madison, Wisconsin.

STYS, S. (1966). *Rekultivace a Tvarovanie Vysypek.* SHD, Most.

TOMAKOV, P.I. (1975). Nekotorie napravlienia po snizieniyu otritzatelnovo vliyania otkrutruch razabotok na okruziaynsyuyu sredu. In *Teoretitieskie i Praktitieskie Problemy Rekultivatsii Narusienych Ziemli.* Moscow.

ZAITZEV, G.A., MOTORINA, L.V. and DANYKO, V.K. (1977). *Land Reclamation for Forestry.* M'isl, Moscow (Russian).

3

Environmental Impacts of Coal Transportation

1 Introduction

WITH THE predicted increase in coal consumption (Chapter 1) it is expected that coal transport activities will increase. It has been suggested, for example, that the U.S.A. will need to invest U.S.$10 billion between 1978 and 1985 in railroads, U.S.$300 million in towboats and barges, and that U.S.$6.4 billion will be required for highway improvements in Appalachia, to meet the increased coal transportation (Morris, Sevian and Crowther, 1979).

The transport of coal represents one component of the complete coal cycle – from exploration and extraction of the fuel, through refining and processing, storage, and finally its conversion to an end-use product for consumers. Increased coal consumption will

TABLE 3.1 *1975 coal transport in the U.S.A. (Morris, Sevian and Crowther, 1979)*

	Tonnage (10^8 tons)*	Average haul distance (km)
Rail	418	480
Waterway	69	765
Truck	79	80–120
Slurry pipeline	10	400

*The studies on which this chapter is based quote tonnages in tons (U.S.): 1 ton (U.S.) = 0.9072 tonne.

involve an increase in activity of each component with consequent environmental impacts.

Coal transportation is executed by rail, truck (lorry), water (on rivers, canals, lakes and intercoastal waterways) and slurry pipeline or conveyer belt. Table 3.1 gives comparative tonnages of coal transported in the U.S.A. and the average haul distance involved for each method of transport. It will be seen that most is transported by rail. However, movement over short distances usually involves trucking, and this is the fastest-growing means of transport. As coal utilization is increased it is envisaged that transport by water will increase, particularly if there is a considerably expanded international trade, and there will also be significant developments in coal movement by slurry pipeline.

Environmental impacts of coal transport occur during *loading, en route* or during *unloading*. The impacts are likely to affect "natural" systems (including agriculture, forestry, horticulture and aquaculture), buildings and installations and involve death or injury to humans in an occupational capacity, or to the public. The environmental impacts on humans have received most attention in a series of studies conducted under contract to the United States Department of Energy (Hamilton, 1979; Morris, Novak and Hamilton, 1979; Morris, Sevian and Crowther, 1979; Szabo, 1978). It is these reports that have been utilized in producing this summary account.

2 Environmental Impacts Common to all Modes of Transport

All forms of transport of coal exhibit certain common environmental impact features. All use land for terminal handling plants or for railroad installations or pipeline throughways. Rail transport and trucks cause damage to buildings and trucking coal causes major structural damage to highways. Air pollutants are emitted from engines powering the transportation facility (train, truck, tugboat, or pump station) and also noise results. Accidents are associated with all forms of transport – usually occupational accidents causing fewer deaths but more injuries, whereas the reverse is true when the public are involved.

The transport of coal in all its forms necessarily involves fugitive dusts, even though precautionary measures are increasingly taken. It is generally estimated that 0.02% of the coal loaded is lost as fugitive dust and a similar percentage lost at unloading. Until recently, loss during transportation has not been estimated but more work is enabling these losses now to be assessed. There is little information of the overall effect of fugitive dust emissions from coal-handling and transport *en route*, although estimates have been made that 16% and 23% of the mean particulate loading of the air in the Atlantic and Mountain regions of the U.S.A. respectively are due to coal handling. It has been estimated that, in the U.S.S.R., air dust content in the vicinity of reloading bays for coal transportation and storage may reach 100 mg m^{-3} and be even higher if low wind velocities prevail. This would lead to effects on human health, crops, other vegetation, animals, and visibility and would prove a general disamenity.

As a rule, occupational impacts occur at loading, unloading and during the movement of coal. Effects to the public mostly occur during coal movement.

3 Rail Transport

3.1 Operational Features

Coal that is to be transported by rail is generally loaded at mine sidings into wagons that will carry 55, 75 or 100 t of coal. The loading and unloading rates, depending on the methods employed, can be several thousand tons per hour. It is not unusual for trains to be assembled to include over 120 wagons, transporting over 11 000 t of coal on a 480 km haul.

3.2 Human Injury and Death

More than half of the accidental deaths and injuries involved in the transport of coal in the U.S.A. result from collisions between trains

TABLE 3.2 *Public and employee deaths and injuries per annum estimated for a 1000 MW(e) power plant*

	Per 10^9 t-miles		Per plant-year		Per train-mile	
	Deaths	Injuries	Deaths	Injuries	Deaths	Injuries
Public	1.74	3.66	1.15	2.4	0.26	0.56
Employees	0.059	4.44	0.04	2.9	0.009	0.68

and motor vehicles at road–rail crossings. Statistics for accidental injury and death are not given according to the freight being carried. Consequently estimates must be obtained for coal transport by assigning a proportionate share based on the weight or volume of coal transported over unit distance (t-miles travelled). Accident rates vary between regions, depending on the density of automobile traffic and the driving patterns. There is thus a considerable degree of uncertainty associated with the derived estimates.

Assumptions have been made in relation to a 1000 MW(e) power plant being fed with coal involving a 480 km coal haul, requiring 2.2×10^6 t of coal a year, equivalent to 6.6×10^8 t-miles. Public and employee death and injury estimates are shown in Table 3.2. It will be seen that there is a far higher public death : injury ratio than there is for employees.

3.3 *Fugitive Emissions*

Fugitive emissions of coal particulates and other air pollutants occur during loading, unloading and *en route*. It is now estimated that about 50% of the coal dust losses occur during journey time and 25% at loading and 25% at unloading. There is some loss also due to spillage. Some estimates suggest that 0.2 kg t^{-1} is emitted during loading and a similar amount during unloading. Losses in transit may be from 0.05 to 1.0% of the total coal involved. For this reason wind guards are used, moistened coal transported or latex polymer or asphalt sprays used. Some wagons incorporate flip-top lids (Szabo, 1978).

Szabo (1978) has given estimates for atmospheric emissions from a unit train carrying 11 430 t of coal making a round trip of 985 km (Table 3.3).

There are no Federal regulations controlling the emission of particulates during coal loading and unloading, and only West Virginia has regulations that are specific to coal handling.

TABLE 3.3 *Atmospheric emissions for a unit train transporting 11 430 t of coal on a round trip of 985 km (Szabo, 1978)*

Emission	kg/trip
Particulates	345
SO_2	780
NO_x	4855
HC	2075
CO	935
Particulates during loading	2285
Particulates during unloading	2285
Fugitive emissions in transit	5700

3.4 Other Effects

Coal transport by rail has a number of other environmental impacts relating to noise (engine, exhaust, horn, wheel–rail interaction and brake cooling blowers). Diesel engines emit exhausts and some chemicals may be leached from the coal during transit.

Sparks from trains may cause bush fires, initially along the permanent way thoroughfare, but in dry regions these may spread to surrounding vegetation. Control measures involve spraying vegetation along the permanent way with chemicals.

4 Coal Transport by Truck

The transport of coal by truck is usually confined to short-distance haulage but is a rapidly growing method of transport. It involves two point crossings (one loaded and one empty) during which exhaust fumes are emitted, spillages occur and fugitive dust losses are experienced (probably 0.04% of the load during combined loading and unloading and 0.05% during transit). For houses and other inhabited buildings near the road, there is exposure to engine, exhaust, tyre and cooling fan noise, sometimes in excess of 100 dB.

It has been estimated (Morris, Novak and Hamilton, 1979) that 7 accidents occur per 10^6 vehicle-miles and there are 0.03 deaths and 0.5 injuries per accident. This is equivalent to 2.4 deaths and 40 injuries for the generation of 1 GW(e) yr^{-1}.

Road damage by haulage trucks is a major environmental cost.

TABLE 3.4 *Deaths and injuries resulting from coal transportation in the U.S.A. (Hamilton, 1979)*

Mode	Accidents per t-mile				Accidents per GW(e) yr^{-1}[b]			
	Public death	Public injury	Occupational death	Occupational injury	Public death	Public injury	Occupational death	Occupational injury
Rail	4.0×10^{-10}	8.6×10^{-10}	1.4×10^{-11}	1.0×10^{-9}	0.41	0.87	0.014	1.1
Truck	7.0×10^{-9}	1.2×10^{-7}	$7.0 - 10^{-9}$	1.2×10^{-7}	1.2	20.0	1.2	20.0
Barge	8.8×10^{-11}	9.0×10^{-11}	2.8×10^{-10}	1.0×10^{-9}	0.14	0.15	0.46	1.6
Pipeline[a]			6.7×10^{-11}	1.1×10^{-10}			0.16	0.26

[a] Coal slurry;
[b] Assumes: 1.0 GW(e) yr^{-1} requires 3.3×10^6 t of coal; coal energy content of 12 000 Btu lb^{-1}; average haul lengths by rail = 300 miles, truck = 50 miles, barge = 450 miles and pipeline = 700 miles.

5 Coal Transport by Water

The transport of coal by water involves barges, loaded by conveyer belt, and assembled for movement by tug, individual motorized barges, or coastal-water ships. The main human hazards are occupational and these are compared with other forms of coal transport in Table 3.4.

In the U.S.A. coal is moved by water on Appalachian rivers, and on the Mississippi and Ohio Rivers. Diesel tugboats, loaded by conveyer belt, transport 18 000 t. During the loading of the amount about 3500 kg of dust would escape. Unloading, by a clamshell grab into a receiving bin, might also result in 3500 kg being lost but this is reduced where receiving bins are equipped with sprays. Estimates of 0.02% over a 2-day trip are given for fugitive escapes during transit.

6 Slurry Pipelines

Although the use of slurry pipelines for coal transport is not greatly developed it is seen as a potential evolution point in coal transport.

The deaths and injuries for slurry pipeline transport have to be estimated from oil pipeline deaths and injuries, taking into account the man-hours required to operate a coal-slurry pipeline. It is regarded as an entirely occupational risk, as shown in Table 3.4.

Slurry pipelines with electrically powered pump stations 100–160 km apart are envisaged. There are no emissions to the air and, due to building insulation and remote siting, noise is not a problem. Water demand is a problem and potential environmental impact as, in the U.S.A., it is not economical to return the water for reuse. At the power plant the water resulting from the dewatering of the slurry has to be disposed of. It is usually saline and loaded with coal "fines".

Any accidental release of coal-slurry from a pipeline has the potential for damage to crops, local vegetation, aquatic systems and aquifers but the operation of pipelines is not yet sufficiently practised to allow estimates of this risk to be given.

7 Coal Transport Impacts in Perspective

It is necessary to keep estimates of the environmental impacts of coal transport in perspective: in relation to other components of the coal fuel cycle; in relation to other means of electricity generation; and in relation to the hazards experienced by human beings in everyday life.

Table 3.5 gives estimates of the number of accidental deaths and injuries attributed to various modes of coal transportation in 1975, in

TABLE 3.5 *Estimated annual accidental deaths and injuries due to coal transportation in the U.S.A. (Hamilton, 1979)*

Transport mode	Subject		Total 1975	1985	1990
Rail	Employee:	deaths	1.7–6.9	3.1–12.3	3.5–14.0
		injuries	130–560	232–996	263–1130
	Public:	deaths	51–219	91–390	103–443
		injuries	108–461	191–820	218–932
Barge	Employee:	deaths	3.3	6.7	8.2
		injuries	321	656	797
Pipeline	Employee:	deaths	0	0.16	0.38
		injuries	0	15	34
Other	Total:	deaths	72	120	130
		injuries	2700	4300	4700

the U.S.A., and as estimated for growth in energy use in 1985 and 1990 according to the conventional mix of energy sources.

In terms of the total fuel cycle it is possible to attribute risk, in terms of deaths, disease and injury to coal mining, coal processing, coal transport and the generation of electricity in coal-fired power stations. These estimates are given in Table 3.6.

TABLE 3.6 *Estimated effects (based on 1000 MW(e) plant-year working at 65% capacity) of the coal fuel cycle (Hamilton, 1979)*

Cycle component	Subject		Deaths	Disease-injury
Mining	Workers:	accidental	0.6	42
		occupational	0.02–0.4	0.5–1.0
Processing	Workers:	accidental	0.05	2.9
		occupational	–	–
Transport	Public + workers:	accidents	0.3–1.3	1.2–5.9
Electricity generation	Public:	air pollution[a]	0.6	no estimate
		air pollution[b]	6	no estimate
	Workers:	accidental	0.1	3.3

[a] Population within 80 km radius of plant.
[b] Total population.

Estimates have compared the total risk to public health of electricity generation to populations within 80 km of plants, for the U.S.A. in 1975 (Hamilton, 1979). This information is given in Chapter 5 (Table 5.8) of this book. It has also been estimated that of the total annual deaths in the U.S.A. between the ages of 1 and 74, 0.2–1.9% are associated with smoking and about 2.5% with car accidents, about half of which are due to drunken drivers.

Estimates derived in the calculation of the environmental hazards of different modes of coal transport are subject to considerable inaccuracies, due to incomplete data and analysis. As coal utilization increases and coal transport with it, these uncertainties will need to be resolved.

8 References

HAMILTON, L.D. (1979). Health effects of electricity generation. *Symposium on Health Effects of Energy Production*, Chalk River Nuclear Laboratories, Ontario, Canada.

MORRIS, S. C., NOVAK, K.M. and HAMILTON, L.D. (1979). *Databook for the Quantitation of Health Effects from Coal Energy Systems.* Biomedical and Environmental Assessment Division, Upton, New York.

MORRIS, S. C., SEVIAN, W. A. and CROWTHER, M. A. (1979). *A Coal Transport Environmental Residuals Model for SEAS.* Biomedical and Environmental Assessment Division, Upton, New York.

SZABO, M.F. (1978). *Environmental Assessment of Coal Transportation.* PEDCo Environmental Inc., Cincinnati, Ohio.

4

Coal Conversion Processes: Gasification and Liquefaction

1 Introduction

There are several factors which make the potential for coal processing plants to discharge hazardous materials different from other fossil or nuclear processing facilities. One of these is the nature of the raw material. Coal has a complex heterocyclic structure which must be broken down to produce desirable products. Cleavage of these cyclic linkages, and the chemical form of the resultant compounds, are sometimes difficult to analyse and predict. In addition, raw coal has less homogeneity than gas or oil in terms of physical structure and the distribution of trace elements. Its structure also presents materials-handling problems, with attendant environmental discharges, not associated with the movements of fluids. For example, coal feeds usually require crushing and grinding into small particles which are then often incorporated into oil or water slurries in order to be uniformly introduced into the liquefaction or gasification process.

Another factor of significant concern is the size of projected coal processing plants. Commercially feasible facilities typically will reduce in excess of 20 000 tonnes of coal per day to products and wastes (NIOSH, 1978b; Battelle, 1973; Ghassemi, Crawford and Quinlivan, 1979). Because of the large amounts of material being processed, the potential exists for environmental discharge of considerable quantities of hazardous materials. These include hydrogen cyanide, phenols, cresols, carbonyl and hydrogen sulphides, ammonia, mercaptans, thiocyanides, aniline, arsenic, trace metals, and various polycyclic hydrocarbons (Schmidt, Sharkey and Friedel, 1974; Moeller, Roberts and Britz, 1974; Cooke and Graham, 1965; Forney *et al.*, 1974; Sexton, 1960a, 1960b; Ketcham and Norton, 1960). In this regard the environmental implications of *in situ* conversion, where even the quantities of coal converted are not well defined, are also of concern. If coal is converted to a secondary energy (gasoline), or feedstocks for chemicals, several process routes are possible, and it seems to be an important aim for future developments to find out the best route with a minimum of pollution. This can be exemplified by looking at the use of coal as

TABLE 4.1 *Energy use for heat: percentage efficiency values and specific emissions for secondary energy options from coal*

	Type of secondary energy			
	Conventional electricity	Coal gas: Town gas	Coal gas: SNG[a]	Coal oil
Primary energy	100	100	100	100
Transport and preparation	98	98	98	98
Primary energy at point of conversion	95	95	95	95
Secondary energy	36	57–69	48–63	48–57
Energy at consumer	33	55–67	46–61	47–56
Useful energy as heat	33	47–57	39–52	38–45
Emission kg $Gcal^{-1}$ useful energy				
Dust	0.5–0.8	0	0	0
SO_x	3.8–5.2	0.04	0.04	0.76
NO_x	Depends on combustion conditions			
CO_2	1220	760	890	960

[a] SNG = Substitute natural gas.

heat supply for space heating and process heat, in the domestic or industrial sectors. For providing heat from coal there exist the routes via electricity production, via gasification and via hydrogenation, and of burning the coal gas or the coal oil at the producers' end. To compare these different routes, both efficiency and specific emissions related to the production of useful heat energy have to be determined for the individual routes. Table 4.1 sets out the efficiency values and specific emissions for various applications of hard coal for producing heat.

The characteristics of each coal conversion process must be considered in any attempt at hazard evaluation, as so many different processes have been attempted or projected (see reviews by Parker and Dykstra, 1978; Ghassemi, Crawford and Quinlivan, 1978; NIOSH, 1978a). Each needs to be analysed specifically for the types of wastes produced and how these are distributed to water, air and soil. However, some general observations can be made about the nature of the conversion process. Most processes require energy addition at elevated temperatures and pressures in order to break

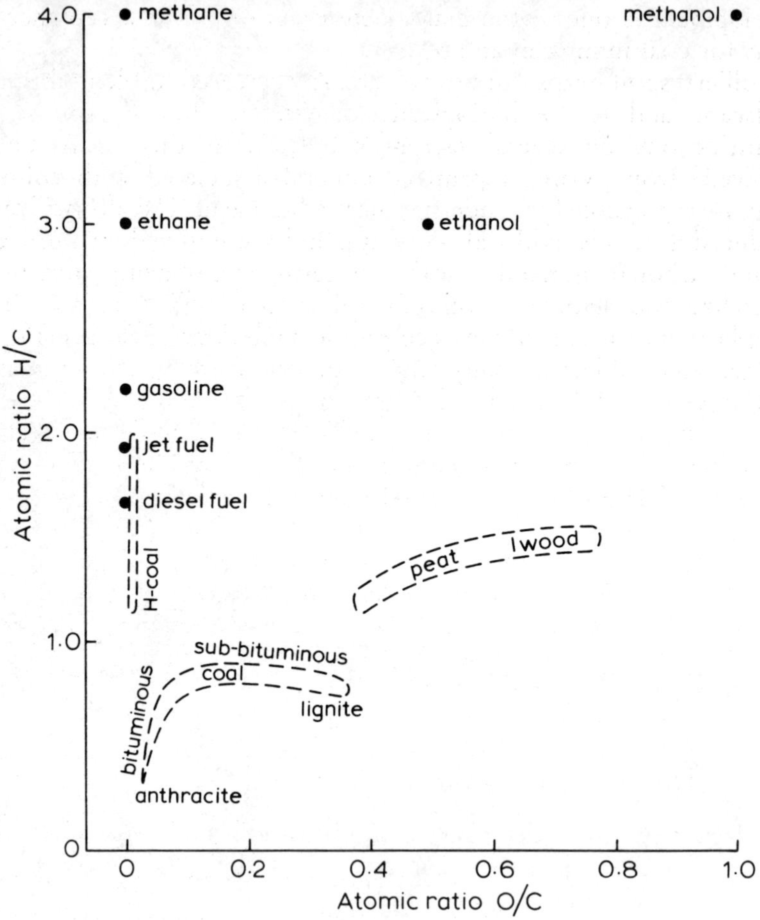

FIGURE 4.1 *The atomic compositions of several raw materials and finished products involved in coal conversion processes*

down the coal. Figure 4.1 indicates the atomic compositions of various raw materials and finished products. The necessary energy is ordinarily derived from oxidising a portion of the incoming coal. Hydrogen is supplied either through direct or indirect utilization of the hydrogen atoms in water, or through use of a hydrogen donor solvent.

Figure 4.2 shows a generalized flow sheet combining a number of processing schemes. The intent of the flow sheet is to emphasize the elements common to most thermochemical conversion processes.

Each particular process is considerably more complex and requires careful analysis of its products and wastes. In general, reaction conditions for gasification are more severe than those for liquefaction, although the unit operations are similar. The gasification processes will be treated in more detail in section 2 and the liquefaction processes in section 3. At present, only gasification processes have been utilized on a large-scale and are operating commercially.

2 Gasification

Gasification is a well-known technique of coal conversion. The first commercial application of modern gasification techniques was in 1926 with the commissioning of a plant using the Winkler process at Leuna. Since the 1930s coal gasification has been applied world-wide for the production of ammonia and hydrocarbons and also for the generation of medium-Btu gas. During the 1960s and 1970s intensive research work has been devoted to the further development of gasification techniques. In the first place this means improving conventional methods with a view to greater economy. To cope with the eventuality of a shortage of natural gas, there are plans to convert coal directly into substitute natural gas (SNG) to be fed into existing natural gas pipeline networks.

An area which appears highly promising, but as yet is in its infancy, is the application of catalysts in coal gasification. Another area of development is the design and operation of combined processes of coal gasification, with subsequent combustion of the coal gas under pressure, for the environmentally acceptable generation of electricity with high thermal efficiencies. Lastly, efforts are being made to reduce the quantities of coal needed for conversion into gas by using process heat from high-temperature nuclear reactors. (A great number of survey articles have been published; von Fredersdorff, Elliott and Lowry, 1963; Jüntgen *et al.*, 1980; von Gratkowski, 1958; Falbe, 1977; Schmidt, 1966; Dolch, 1936; Schilling, Bonn and Krauss, 1979; VGB Konferenz, 1979; Elgin, 1976; Robson, 1977; Jüntgen, 1977, 1979b; van Heek, 1978; Peters and Schilling, 1977; Schulze, 1974, 1978).

2.1 Basic Principles

Coal gasification means the complete conversion of coal into gas using heterogeneous gas solid reactions. The main process is the reaction of the carbon of coal with steam, at pressures below 100 bar

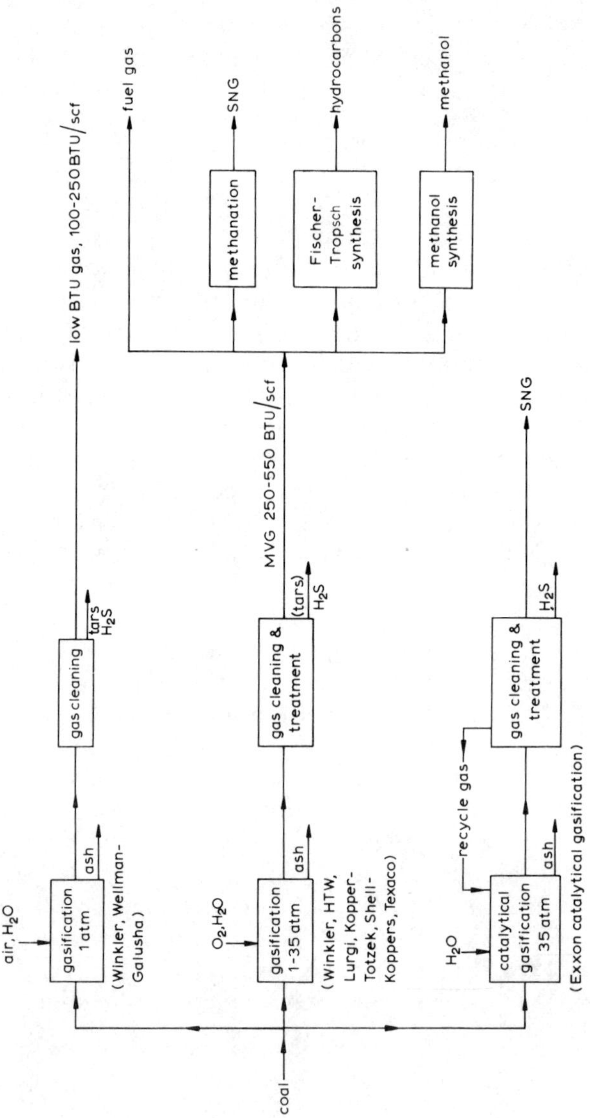

(a)

FIGURE 4.2 *Generalized flow diagram combining a number of coal conversion processes (Considine, 1977; Hoffman, 1978; Wadden, 1976; Bodle and Vyas, 1975)*

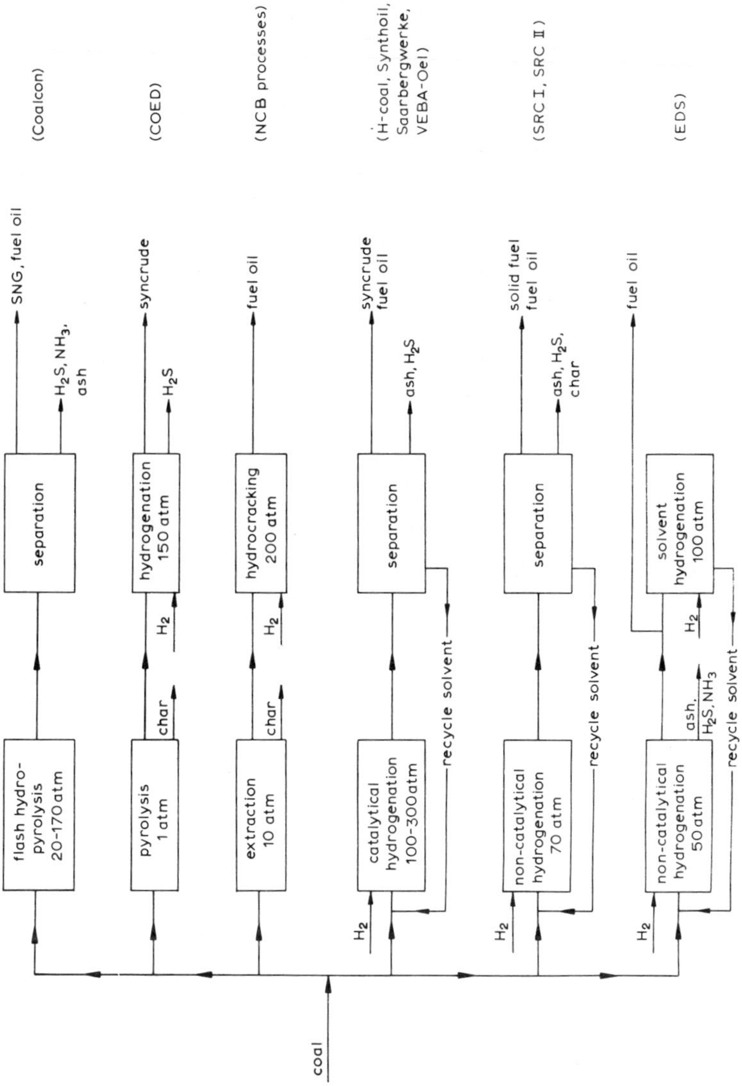

(b)

FIGURE 4.2 *continued*

and temperatures above 750°C, to form a "synthesis gas" containing mainly CO and H_2 with smaller amounts of CO_2 and CH_4 depending on process conditions. To perform this strongly endothermal reaction in technically feasible systems the introduction of process heat is necessary, which is possible by transfer of heat from outside (allothermal processes) or by burning part of the coal inside the reactor using oxygen (autothermic processes). This latter process is the basis for the production of hydrogen, ammonia, methanol, hydrocarbons using the Fischer-Tropsch synthesis and methane using the so-called catalytic methanation reaction. Another method for converting coal into a gas is the direct exothermal conversion of coal with hydrogen into methane at a similar range of pressure and temperatures as mentioned above. This reaction can also be combined with the steam carbon reaction and in this way leads to a direct conversion of coal into SNG.

Coal contains not only carbon but also volatile matter, which is released during the heating up of coal in the gasification reactor by the so-called pyrolysis reactions. The interaction of pyrolysis and gasification of coal is mainly dependent on grain size and rate of heating of the feed coal. At low rates of heating the pyrolysis products are released first and the remaining char reacts with the gasification agent later on at high temperatures, yielding gaseous products from the carbon. In this case, if a countercurrent of gas and solids in the gasification reactor is induced, the product gas contains pyrolysis products such as tar, phenols and higher hydrocarbons. At high rates of heating, the pyrolysis reaction is shifted to higher temperatures and in this way both – pyrolysis products at the outer grain shell and carbon – react with the gasifying agent to form the gasification products H_2, CO, CO_2, CH_4 and less or no pyrolysis products are found in the product gas.

The nature of the reaction products of coal gasification is primarily determined by the chemical equilibria which are established, especially at high temperatures, and sufficient residence times of coal and products in the gasification reactor. Thus a high pressure and a low temperature increase the methane yield and decrease the hydrogen and the carbon monoxide yield. Much information exists on this topic (Hedden, 1962; Hedden and Mienkina, 1965; Peters, 1976; van Heek, Jüntgen and Peters, 1973; Feistel, van Heek and Jüntgen, 1978; Yoon, Wei and Denn, 1978; Jüntgen and van Heek, 1979a; Jüntgen, 1979a).

TABLE 4.2 *Typical commercial gasification processes*

	Operating principle		
	Moving bed	Fluidized bed	Entrained bed
Typical commercial process	Lurgi pressure gasification	Winkler process	Koppers-Totzek process
Reaction conditions	P up to 35 bar T Dry pyrolysis 400°C T Gasification 800–1000°C T Combustion 1000°C	P = 1 bar T = 1000°C	P = 1 bar T = 1800°C
Coal type	Lignite Medium-caking bituminous	Lignite Low-caking bituminous coal	Lignite Bituminous coal
Particle size of coal	Relatively large; optimum size range 6–40 mm	Large to medium; optimum size size range 0–8 mm	Fine crushing required; 80–90% <0.1 mm
Ash removal	Dry	Dry	Slag
Steam consumption	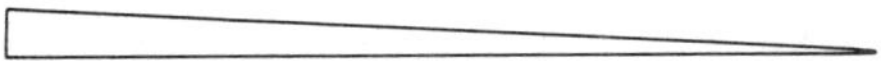		
Oxygen consumption	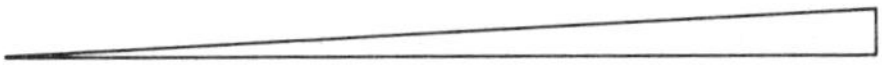		
Gas composition CH_4	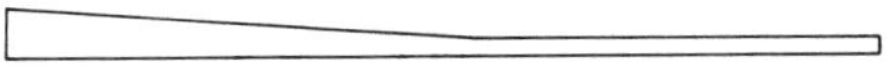		
CO	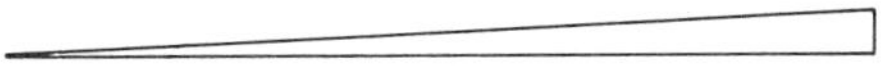		
Tars, phenols	+	–	–
Gas production per gasifier	46 000 $Nm^3 h^{-1}$ raw gas 30 000 $Nm^3 h^{-1}$ purified gas (Sasol II)	17 000 $Nm^3 h^{-1}$ 20 000 $Nm^3 h^{-1}$	20 000 $Nm^3 h^{-1}$ (2 burners) 50 000 $Nm^3 h^{-1}$ (4 burners)
Application	Town gas production Synthesis gas for F–T synthesis + CH_3OH synthesis	Synthesis gas for chemical industry	Synthesis gas for NH_3 production

TABLE 4.3 *Dry ash Lurgi commercial installations using the West German Lurgi technology*[a] *(Ghassemi, Crawford and Quinlivan, 1979)*

Plant No.	Location	Year	Type of coal	Gasifier ID (m)	Capacity (Nm^3 day^{-1} $\times 10^6$)	No. of gasifiers
1	Bohlen, Central Germany	1940	Lignite	2.6	0.25	5
2	Bohlen, Central Germany	1943	Lignite	2.6	0.28	5
3	Most, C.S.S.R.	1944	Lignite	2.6	0.21	3
4	Zaluzi-Most, C.S.S.R.	1949	Lignite	2.6	0.25	3
5	Sasolburg, South Africa	1954	Sub-bituminous with 30% ash and more	3.7	4.20	9
6	Dorsten, West Germany	1955	Caking sub-bituminous with high chlorine content	2.7	1.54	6
7	Morwell, Australia	1955	Lignite	2.7	0.62	6
8	Daud Khel, Pakistan	1957	High volatile coal with high sulphur content	2.7	0.14	2
9	Sasolburg, South Africa	1958	Sub-bituminous with 30% ash and more	3.7	0.53	1

10	Westfield, Great Britain	1960	Weakly caking sub-bituminous	2.7	0.78	3
11	Jealgora, India		Different grades	not available	0.03	1
12	Westfield, Great Britain	1962	Weakly caking sub-bituminous	2.7	0.25	1
13	Coleshill, Great Britain	1963	Caking sub-bituminous with high chlorine content	2.7	1.30	5
14	Naju, Korea	1966	Sub-bituminous with 30%	3.7	2.10	3
15	Sasolburg, South Africa	1966	Sub-bituminous with 30% ash and more	3.7	2.10	3
16	Luenen, F.R.G.	1970	Sub-bituminous	3.4	1400×10^6 Btu hr^{-1}	5
17	Sasolburg, South Africa	1973	Sub-bituminous with 30% ash and more	3.7	5.32	3
18	Sasolburg, South Africa	1978	Sub-bituminous with 30% ash and more	4.0	42	36

[a] A parallel development of the Lurgi process has been carried out by the German Democratic Republic since the Second World War and several commercial plants using the East German technology currently exist in Eastern European countries. Examples include the Gaskombinant Schwarze Pumpe, East Germany with 24 gasifiers (3.6 m ID)

2.2 *Commercial Processes*

Table 4.2 outlines the gasification processes most commonly used and Tables 4.3 and 4.4 list some of the full-scale gasification plants that have been built since 1940. The moving-bed type reactor is used in the Lurgi pressure gasification process – the only process which has so far been carried out using high pressure – the fluidized-bed principle is used in the Winkler process and the entrained-bed type reactor in the Koppers-Totzek process. These processes are classified according to the conditions of the reaction, which leads to very different temperatures in the gasifiers. As a result the throughput rate per unit volume of the reactor increases with temperature but does not allow any assessment of the processes. In the moving bed, changes in temperature and pressure have a considerable influence on the development of the reaction. In the Winkler gasifier the influence of pressure and temperature is less marked. In the Koppers-Totzek gasifier, increases in temperature have hardly any influence.

The Lurgi pressure gasification method has proved exceptionally versatile in its applications to date. In particular, large quantities of synthesis gas are produced world-wide for Fischer-Tropsch synthesis. A contract has also been concluded for the production of synthesis gas for methanol. There are also plants producing gas of town-gas quality. The Winkler method is generally used for producing synthesis gas in the chemical industries, and is operated in the CMEA countries predominantly on lignite. The Koppers-Totzek process is used all over the world with widely differing types of coal wherever gas has to be produced for the synthesis of ammonia. The modern plants are geared to an ammonia production capacity of 1000 tonnes day^{-1}. A number of articles have been published on special gasification processes and are itemized appropriately here (*Moving bed*: Hiller, 1975; Hoogendorn, 1976; Röbke, 1978; *Fluidized bed*: Flesch and Velling, 1962; Davy Powergas, 1973; Banichek, 1973; Franke, 1978; Anwer and Bögner, 1976; *Entrained bed*: Göke, 1979; Linke and Vogt, 1979; Völkel *et al.*, 1979; Franzen, 1977; Staege, 1976; Seipenbusch and Ruprecht, 1978; Kraaijveld, 1978; Rossbach *et al.*, 1978; *Use of nuclear reactor*: Arndt *et al.*, 1979; Jüntgen and van Heek, 1977, 1979b).

TABLE 4.4 *Gasification plants using the Koppers-Totzek process (Ghassemi, Crawford and Quinlivan, 1978)*

Location	Fuel	Number of gasifier units	Capacity ($CO + H_2$ in 24 hr)	Use of synthesis gas	Year of order
Charbonnages de France, Paris, Mazingarbe Works (P.d.C.) France	Coal dust, coke oven gas, tail gas	1	75 000–150 000 Nm^3 2 790 000–5 580 000 SCF	Methanol and ammonia synthesis	1949
Typpi Oy, Oulu, Finland	Coal dust, oil, peat	3	140 000 Nm^3 5 210 000 SCF	Ammonia synthesis	1950
Nihon Suiso Kaisha, Ltd, Tokyo, Japan	Coal dust	3	210 000 Nm^3 7 820 000 SCF	Ammonia synthesis	1954
Empresa Nacional "Calvo Sotelo" de Combustible Liquidos y Lubricantes, S.A., Madrid, nitrogen works in Puentes de Garcia Rodriquez, Corunna, Spain	Lignite dust	3	242 000 Nm^3	Ammonia synthesis	1954
Typpi Oy, Oulu, Finland	Coal dust, oil, peat	2	140 000 Nm^3 5 210 000 SCF	Ammonia synthesis	1955
S.A. Union Chimique Belge, Brussels, Zandvoorde Works, Belgium	Bunker-C-oil plant convertible for coal dust gasification	2	176 000 Nm^3 6 550 000 SCF	Ammonia synthesis	1955
Amoniaco Portugues S.A.R.L., Lisbon, Estarreja Plant, Portugal	Heavy gasoline, plant extendable to lignite and anthracite dust	2	169 000 Nm^3 6 300 000 SCF	Ammonia synthesis	1956
The Government of Greece; The Ministry of Coordination, Athens; nitrogenous fertilizer plant, Ptolemais, Greece	Lignite dust, bunker-C-oil	4	629 000 Nm^3 23 450 000 SCF	Ammonia synthesis	1959

TABLE 4.4 *Continued*

Location	Fuel	Number of gasifier units	Capacity ($CO + H_2$ in 24 hr)	Use of synthesis gas	Year of order
Empresa Nacional "Calvo Sotelo" de Combustibles Liquidos y Lubricantes, S.A., Madrid, nitrogen works in Puentes de Garcia Rodriquez, Corunna, Spain	Lignite dust or naphtha	1	175 000 Nm^3 6 500 000 SCF	Ammonia synthesis	1961
The General Organization for Executing the Five Year Industrial Plan, Cairo; nitrogen works of Societé el Nasr d'Engrais et Chimiques, Attaka, Suez, U.A.R.	Refinery off-gas, LPG, and light naphtha	3	778 000 Nm^3 28 950 000 SCF	Ammonia synthesis	1963
Chemical Fertilizer Company Ltd, Thailand, synthetic fertilizer plant at Mao Moh, Lampang, Thailand	Lignite dust	1	217 000 Nm^3 8 070 000 SCF	Ammonia synthesis	
Azot Sanayii T.A.S., Ankara, Kutahya Works, Turkey	Lignite dust	4	775 000 Nm^3 28 850 000 SCF	Ammonia synthesis	1966
Chemieanlagen Export–Import GmbH, Berlin für VEB Germania, Chemieanlagen and Apparatebau, Karl-Marx-Stadt, VEB Zietz Works	Vacuum residue and/or fuel oil	2	360 000 Nm^3 13 400 000 SCF	Raw gas to produce hydrogen for hydrogen generation	1966
Kobe Steel Ltd, Kobe, Japan, for Industrial Development Corp., Zambia, at Kafue near Lusaka, Zambia, Africa	Coal dust	1	214 320 Nm^3 7 980 000 SCF	Ammonia synthesis	1967

Nitrogenous Fertilizers Industry S.A., Athens, nitrogenous fertilizer plant, Ptolemais, Greece	Lignite dust	1	165 000 Nm^3 6 150 000 SCF	Ammonia synthesis	1969
The Fertilizer Corporation of India Ltd, New Delhi, Ramagundam Plant, India	Coal dust	4 (1 of them as standby)	2 000 000 Nm^3 74 450 000 SCF	Ammonia synthesis	1969
The Fertilizer Corporation of India Ltd, New Delhi, Talcher Plant, India	Coal dust	4 (1 of them as standby)	2 000 000 Nm^3 74 450 000 SCF	Ammonia synthesis	1970
Nitrogenous Fertilizers Industry S.A., Athens, nitrogenous fertilizer plant, Ptolemais, Greece	Lignite dust	1	242 000 Nm^3 9 009 000 SCF	Ammonia synthesis	1970
The Fertilizer Corporation of India Ltd., New Delhi, Korba Plant, India	Coal dust	4 (1 of them as standby)	2 000 000 Nm^3 74 450 000 SCF	Ammonia synthesis	1972
AE & CI Ltd, Johannesburg, Modderfontein Plant, South Africa	Coal dust	6	2 150 000 Nm^3 80 025 000 SCF	Ammonia synthesis	1972
Indeco Chemicals Ltd, Lusaka, Kafue Works, Zambia	Coal dust	1	220 800 Nm^3 8 220 000 SCF	Ammonia synthesis	1974
Indeco Chemicals Ltd, Lusaka, Kafue Works, Zambia	Coal dust	2	441 660 Nm^3 16 440 000 SCF	Ammonia synthesis	1975

SCF = Standard cubic feet
Nm^3 = Normal cubic metre (0°C and 1 atm)

2.3 Development Trends

2.3.1 Improved processes

Although the gasification processes described are being applied successfully world-wide for the industrial production of gas from coal, they still need to be improved to make them more economic. There is also a need to extend the scope of fluidized-bed technology to include different coal types. Individual research projects in this field are in progress, notably in the U.S.A., the U.K., and the F.R.G., these are listed in Table 4.5. They are categorized according to the different measures introduced to improve techniques.

2.3.2 New processes

Use of catalysts

If it proved possible to generalize the use of catalysts, considerable advantages could be expected for the implementation of the gasification processes. For example, increasing the rate of reaction would lead to a bigger unit throughput per reactor and the lowering of the gasification temperature would result in energy savings, fewer materials problems and higher methane concentrations in the product gas. Moreover, the heat from the hydrogasification could be used directly for endothermic steam gasification, obviating the need to use oxygen for heat generation in autothermic gasification processes.

At present, however, there are still problems with their application on a commercial scale since this requires the use of large quantities of catalyst, and this means extra cost. Also, most catalysts cause appreciable corrosion and are extremely difficult to recover from the coal ash.

Use of nuclear reactor

In autothermic coal gasification the necessary reaction heat has to be generated by a partial combustion of coal with oxygen, which limits the maximum gasification efficiency to between 60 and 70%. A reduction in the specific coal consumption can only be achieved by providing the necessary process heat from other energy sources. One possible source is the high-temperature nuclear reactor with a helium outlet temperature of 950°C. This technique offers advantages over

TABLE 4.5 *Industrial development projects catagorized by the different measures to improve techniques (t = tonnes)*

Measures	Increased gasification temperature	Increased pressure	Increased pressure and temperature	Increased pressure to 100 bar. Change in procedure	Gasification system 2-stage
Anticipated advantages	Increased throughput and increased steam decomposition	No need for compression before gas preparation	Increased throughput	Increased CH_4 content. Improvement in gas preparation	Use of caking coals
Examples of process development	Lurgi "slagging gasifier"	Shell-Koppers Saarberg-Otto Texaco	HT-Winkler	Lurgi pressure gasification	Synthane Bigas
Techniques applied	Gravitating bed Liquid slag removal	Pulverized coal burners Burners with a coal suspension feed	Fluidized bed	Gravitating bed Separate gas removal	Combination of fluidized and entrained bed
Status	Operational (bituminous coal)	Commissioned 6 t hr^{-1} Underconstr. 11 t hr^{-1} Commissioned 6 t hr^{-1} (bituminous coal)	Commissioned 1 t hr^{-1} (Dry lignite)	Under construction 7 t hr^{-1}	Operational 3.1 t hr^{-1} Operational 10 t hr^{-1} (bituminous coal)

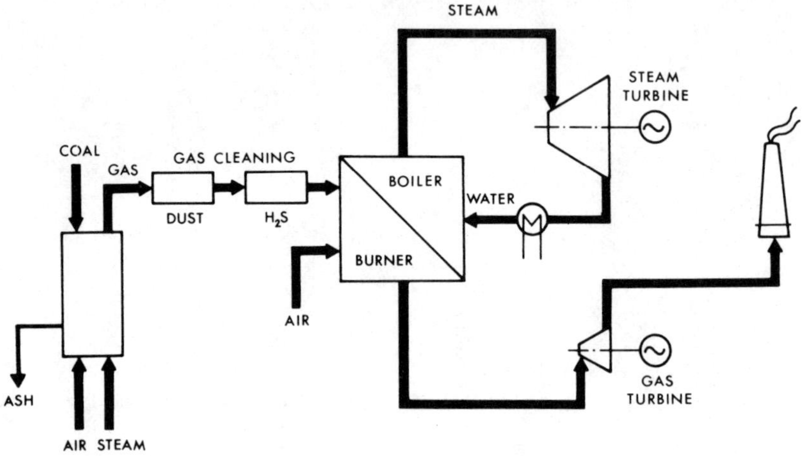

FIGURE 4.3 *Simplified flow diagram of combined gasification and electricity generation*

autothermic coal gasification processes, including the generation of more gas per unit of coal, reduction in size of the gas preparation plants, together with greater economy of the process when more expensive European coals are used.

Use of combined cycles

Since tighter environmental legislation makes the conventional coal-based processes of electricity generation more expensive, investigations have been made to establish whether a non-polluting electricity generation process can be developed based on a combination of gasification and combustion of the coal gas. The following advantages were identified: the use of an additional gas turbine means that this process is likely to be more efficient than conventional methods of electricity generation; the dust and sulphur are removed from a relatively smaller quantity of gas before combustion, in contrast to the conventional processes where flue gas desulphurization takes place downstream of combustion; there is potential for using proven methods of H_2S and dust removal in cases where the process gas is cooled before combustion; the process is generally less polluting in terms of dust, sulphur and NO_x emissions than conventional processes of electricity generation.

A flow diagram of the type of pressure gasification process used by Steag is shown in Figure 4.3.

2.4 Emission Sources of Gasification Plants

Dust emissions may be generated during stocking, handling, crushing, grinding and screening of the raw coal, and handling or stocking the ash produced in the plants. These problems are well-defined in all coal conversion plants and not specific for gasification plants. There are accepted measures for keeping pollution below the statutory limits. Dust-generating equipment has to be housed in dust-tight casings with suction facilities for dust-laden air. This dust-laden air is then treated in special separation units comprising cyclones, bag-type filters, electrostatic precipitators or wet dust separation systems. Also prevention measures against emission risks during stocking are possible, such as spraying with water or chemicals, partial or total tree planting or the provision of shelter.

Differences in the pollution control of different coal gasification processes can arise from the different grain size of coal needed in the gasifier. Both secondary energy conversion and heat production emit dust, SO_2, NO_x and CO_2. As far as SO_2 is concerned, emissions are more serious in coal-based electricity generating, despite stack gas desulphurization, than is the case with the use of either coal oil or coal gas for heating purposes. Coal gas meets particularly well the demands of pollution control in regard to sulphur emissions, requiring that the sulphur (as converted into H_2S) has to be removed virtually completely. This is standard practice. NO_x emissions cannot be fully assessed yet as they are governed more by the firing conditions than by the type of fuel. When conventional energy sources are used to produce heat, CO_2 is inevitably emitted and the CO_2 emissions are inversely proportional to the cumulative efficiency pattern. Emissions are thus at their lowest with gasification, but reach their peak when coal is used to generate electricity.

Other sources of emission in gasification plants are off-gas streams of different stages which depend on the special gasification process and on the design of special gas purification systems used. They are exemplified in Chapter 8 by discussion of a Lurgi plant.

Finally, the properties of the residual coal ash depend on the gas atmosphere and on the maximum temperature in the gasifier. Therefore, questions of the distribution of trace elements and disposal of gasifier ash can only be discussed considering both the raw coal used and the temperature-time history of the solids moving through the gaseous atmosphere of the gasifier. In this context it should be stressed that only part of the trace elements in coal ash is environmentally relevant: that part which is soluble in water under particular environmental conditions. On the other hand, it must be expected that due to the reducing atmosphere in the gasifier, a part of the trace element content is converted into volatile products such as hydrides, carbonyls or sulphides and in this way is entrained in the raw gas. Therefore, there may be enrichment of the dust removed from the raw gas downstream of the gasifier, and in the case of a low-temperature gasification process this will occur with the tar. If the dust removal is by wet scrubbing, trace elements may also be found in the waste water. A great number of articles have been published on the subject of pollution control related to gasification (Magee, Jahnig and Shaw, 1974; Shaw and Magee, 1974; Jahnig, 1975a, 1975b; Hall, 1974; Jüntgen and Klein, 1976; Beckner, 1975; Massey and Dunlop, 1976; Becker, 1979; Rolke, 1979).

3 Liquefaction

There are three major considerations that have made coal liquefaction an attractive proposition:

(1) the energy-consuming pattern in many countries is optimized for oil as an energy raw material; the most appropriate way therefore to introduce coal is to convert it to an oil; this implies adding hydrogen to increase the H : C ratio and removing oxygen (Figure 4.1).
(2) sulphur and nitrogen in coal yield noxious emissions when coal is burnt; liquefaction by hydrogenation removes sulphur and also much of the nitrogen.
(3) liquefaction of coal makes it easy to separate inorganic, often sulphur-rich, fraction (e.g. by filtration).

3.1 Direct Liquefaction

Solid coal can be converted directly to a liquid by thermo-chemical treatment, at high temperature and high pressure. This route is called direct liquefaction in contrast to indirect liquefaction involving coal gasification to synthesis gas and subsequent synthesis to liquids.

3.1.1 Basic principles for and development of direct liquefaction processes

Chemistry

Coal is made up of a broad spectrum of molecules with a high molecular weight, these are called clusters and are partially cross-linked. The heteroatoms O, S and N are often present in heterocyclic structures. Schematically, the mechanism for liquefaction consists of the following steps:

(1) pyrolysis at about 450°C to break the C–C and the C–O bonds between clusters, resulting in formation of gases, liquids and free radical intermediates;
(2) free radical recombination to form coke, or if hydrogen is present at a sufficiently high concentration, radical transformation to compounds rich in hydrogen;
(3) reaction of the hydrogen to form H_2S, NH_3 and H_2O, which explains the reduction in heteroatom concentration in the product.

Both the pyrolytic cracking and the hydrogenation reactions can be catalysed. Sn, Fe, Mo, Co, Zn and Ni compounds have been used as catalysts. The hydrogen consumption is in the range 2–11% of the coal weight. The more hydrogen that is added, the lighter is the product achieved.

Technology

From a technical point of view, liquefaction generally involves the following procedures: hydrogen production, coal hydrogenation and product refining.

The key step is the hydrogenation and the many processes proposed for direct liquefaction differ in the way this is carried out. In the direct liquefaction processes, hydrogenation is always carried out with the coal in a slurry, acting as a heat carrier and solvent for hydrogen and the subsequent products. The choice of a proper solvent for the slurry preparation is crucial to the process. The process usually operates at a high pressure (100–700 atm) when molecular hydrogen is used for the hydrogenation. A more moderate pressure (about 10 atm) is sufficient when a donor solvent is used instead. A typical residence time for the coal in the chemical reactor is 1–2 h. The use of catalysts in the hydrogenation step allows a lower pressure and shorter residence times.

The hydrogen production is an important step from a financial point of view. It accounts for as much as 30% of the net investment in, for example, an H-coal plant. Unreacted carbon and product gases from the hydrogenation step are used as hydrogen sources. The char is gasified (e.g. in a Texaco generator) and subsequent gas treatment (shift – CO_2 removal) results in hydrogen production.

Product refining always starts with the separation of ash and unreacted coal. Several techniques are used for this: filtration; centrifugation (hydrocycloning); precipitation with an appropriate solvent; distillation. The raw liquid product oil is often further hydrotreated to yield a lighter product suitable for transportation purposes.

Recent developments of the direct liquefaction techniques in different countries are presented briefly in sections 3.1.2–3.1.6, giving examples of different projects.

3.1.2 Developments in the Federal Republic of Germany

Saarbergwerke AG process

This is a modified Bergius process. An iron catalyst is used and a recirculated, refined liquid product is used as solvent. The operating pressure is <300 atm and the hydrogen consumption 5–7%. A 6 t day^{-1} plant was started at Voelklingen in 1980. A 2×10^6 t yr^{-1} plant is planned for start-up in 1987 (Peters, 1979; Strobel, Romey and Koelling, 1979; Würfel and Jorzyk, 1979).

VEBA-Oel AG/Ruhrkohle AG process

Based on the experiences from a 0.5 t day^{-1} liquefaction unit at Bergbau Forschung GmbH, Essen, a 200 t day^{-1} unit has been built at Bottrop. Start-up was planned for early in 1981. The process operates at 250–300 atm, uses an iron catalyst and will have a hydrogen consumption of 6%. A 6×10^6 t yr^{-1} plant is planned for 1987 (Friedrich, Romey and Strobel, 1980; Langhoff *et al.*, 1979; Peters, 1979).

3.1.3 Developments in Great Britain

NCB supercritical process

A 0.1 t day^{-1} unit for supercritical extraction (e.g. with toluene) at 100–200 atm and 360°C has been operated since 1977. The extract is upgraded by hydrocracking after removal of the solvent. A 24 t day^{-1} unit is planned.

NCB solvent extraction process

A 0.75 t day^{-1} unit for extraction of coal with a process-derived solvent at 10 atm and 400°C has been operated since 1977. The extract is hydrocracked at 200–300 atm and 470°C. A 24 t day^{-1} unit is planned. The hydrogen consumption is 5%. Details are given by Urquhart, Martin and Whitehead (1979) and Topper (1980).

3.1.4 Developments in Poland

The technology for recovering liquid fuel from coal developed in Poland is based on extraction of coal and hydrogenation of the extract. At the Carbochemistry Institute of the Central Mining Institute at Tychy-Wyry a 1.2 t day^{-1} unit has been operated since 1977. A nickel/molybdenum catalyst and recycled oil as a donor solvent are used. The operating pressure is 240–250 atm. Research is currently being carried out of pyrolysis using fluidized-bed conditions on pyrolysis by using different kinds of heat carriers. At this Institute tests have been performed to hydrogenate tars (Hulisz *et al.*, 1979; Ihnatowicz, 1979).

3.1.5 Developments in the U.S.A.

Exxon Donor Solvent

A donor solvent process has been developed by the Exxon company at Baytown, Texas. The hydrogenation of the coal is performed by using a donor solvent and hydrogen at 50 atm and 430°C in a plug-flow reactor. No catalyst is employed in the coal hydrogenation step. The donor solvent is regenerated catalytically (nickel/molybdenum) by hydrogenation at 90 atm and 440°C. The process development has proceeded from a 0.5 t day^{-1} unit, started in 1972, to a 1 t day^{-1} unit started in 1974, and finally a 250 t day^{-1} pilot plant which started up in June 1980 (Batchelor, 1979; Epperly and Tauntan, 1979).

H-coal

The Hydrocarbon Research Company, Trenton, New Jersey, has been developing a process based on catalytical (cobalt/molybdenum) hydrogenation of a slurry (mixed raw product and coal), in a fluidized bed at 440°C and 100 atm. The sulphur-resistant catalyst is fluidized by the slurry, and the reactor design makes it possible to change the catalyst continuously. The hydrogen consumption is 5%. A 3 t day^{-1} unit has been operated since 1969. A 600 t day^{-1} unit was started in May 1980 in Catlettsburg, Kentucky (Batchelor, 1979; NAS, 1977; De Vaux and Dutkiewicz, 1979; Wolk, 1979; Stotler, MacArthur and Comolli, 1980).

SRC I and SRC II

Solvent-refined coal processes have been developed in two steps. SRC I operates at 70 atm and 425°C, with a process-derived solvent. The hydrogen consumption is 3% and the product is mainly a low-sulphur solid fuel. SRC II is a modification of SRC I to make it possible to produce a liquid fuel. The process operates under similar conditions but with a more refined solvent. The hydrogen consumption is 4%. None of the processes requires the addition of a catalyst, but on the other hand the processes operate well only with high-ash bituminous coals with a high content of volatiles.

A 6 t day^{-1} SRC I unit has been operated since 1973 in Wilsonville, Alabama, and a plant scaled-up to 6000 t day^{-1} was planned at Newman, Kentucky. The SRC II process has been tested in a 30 t day^{-1} unit since 1976 at Fort Lewis, Washington (a converted SRC I unit). This was planned to be scaled-up to 6000 t day^{-1}. The full-scale unit will be located in Morgantown, West Virginia. The start-up date for the two 6000 t day^{-1} units was set for October 1984 (Urquhart, Martin and Whitehead, 1979; Koenig and Schmalzer, 1979; Moschitto, 1978).

Synthoil

The Synthoil process is based on the use of a fixed-bed catalytic hydrogenation. The U.S. Bureau of Mines started the development in 1970 and in 1978, a 10 t day^{-1} unit was started up. The operation has now been suspended (Nowacki, 1979; Rogers *et al.*, 1978).

3.1.6 Developments in the U.S.S.R.

The Soviet Combustible Fuels Institute has been developing a process for coal liquefaction since 1960. A mixture of high-sulphur crude oil, from which the heavy fraction has been removed, is used as solvent. The hydrogenation is performed at 100 atm and 430°C using an iron/molybdenum catalyst. A method for recovery of the molybdenum has also been developed. A further improvement of the technique and a scaled-up revision is planned (Kriczko and Julin, 1973; Kriczko, 1977).

3.1.7 Future development

The future development of the direct liquefaction technique will probably not lead to much-increased efficiency, but with increasing oil prices and oil shortage the processes will be more attractive financially. Full-scale plants are planned today in the U.S.A., the F.R.G. and South Africa.

3.2 Indirect Liquefaction

The indirect route from coal to liquid fuels is at present the only one in full-scale operation. The technique involves total gasification of the coal to a raw gas according to the well-proven Lurgi or Koppers-Totzek methods. The raw gas is then refined to a synthesis gas which is catalytically converted to a liquid fuel. The conversion method is either the Fischer-Tropsch synthesis, yielding hydrocarbons, or to a lesser extent methanol synthesis. Methanol can be, in its turn, used as raw material for the Mobil process to give hydrocarbons. The objective is to produce a transportation fuel (or a chemical product).

Methanol

Methanol synthesis is based on early work by the German company BASF on the reaction between carbon monoxide and hydrogen. In 1923 it was found that methanol synthesis is selective when a $ZnO–Cr_2O_3$ catalyst is used. A very high pressure, 200–400 atm, was necessary. During the 1960s ICI developed a Cu–ZnO catalyst that made it possible to reduce the pressure to 50–100 atm. Today the technique of making methanol from a synthesis gas is well developed and at least five processes are commercially available.

The Mobil Oil Company has developed a zeolite catalyst that is capable of converting methanol to hydrocarbons with a very high efficiency. This process has been tested in a pilot plant and a full-scale plant is under construction in New Zealand.

Fischer-Tropsch process

The Fischer-Tropsch technology today exists in two versions:

(1) the Arge process yielding straight-chain hydrocarbons suitable for diesel engines;
(2) the Synthol process yielding branched and olefinic hydrocarbons suitable for Otto engines.

Future processes

Much world-wide research is now devoted to finding new catalysts for the conversion of synthesis gas to liquid fuels and other chemical products. French work on catalytic conversion to ethanol, for instance, appears promising.

3.3 Emission Sources of Liquefaction Plants

In principle, the process of coal liquefaction involves chemical compounds which also occur in other carbochemical or petrochemical processes. It is well known that a number of these substances may be carcinogenic, teratogenic or mutagenic. The concentration of benzo(α)pyrene in coal liquefaction products can be between 50 and 60 $\mu g\ g^{-1}$. The concentration in these coal products may be 10–100 times higher than that in cigarette smoke (Budden and Zieger, 1978). Various substances introduced during liquefaction processes (certain catalysts, inhibitors and antioxidants) are also health hazards. Before liquefaction of coal is practised widely on a commercial scale toxicological evaluation needs to be implemented.

The emissions from a liquefaction plant may be more severe than from a gasification plant due to the higher complexity in the former. The presentation of a generalized liquefaction process in Figure 4.2 is very simplified and Figure 4.4 gives a more representative picture of the complexity. A liquefaction plant also contains a gasification plant.

The continuous emissions to the atmosphere from a liquefaction plant will mainly come from the auxiliary facilities, the sulphur recovery section, the power and steam generation facilities and from the drying unit.

Liquid emissions are present in waste-water streams. The effluent is rather similar to that from coke ovens. The effluents from the Polish 1.2 t day^{-1} liquefaction unit represents 6–7% of the coal by

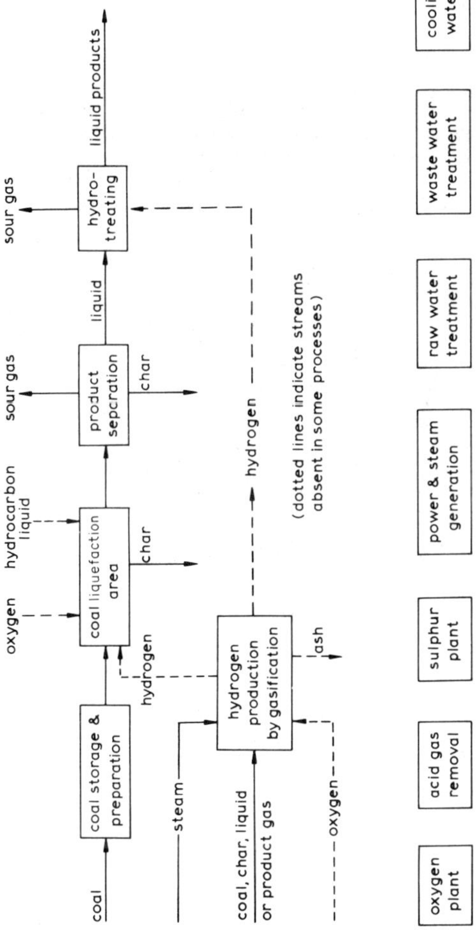

FIGURE 4.4 *Flow diagram representing a generalized liquefaction process (DOE/HEW/EPA, 1978)*

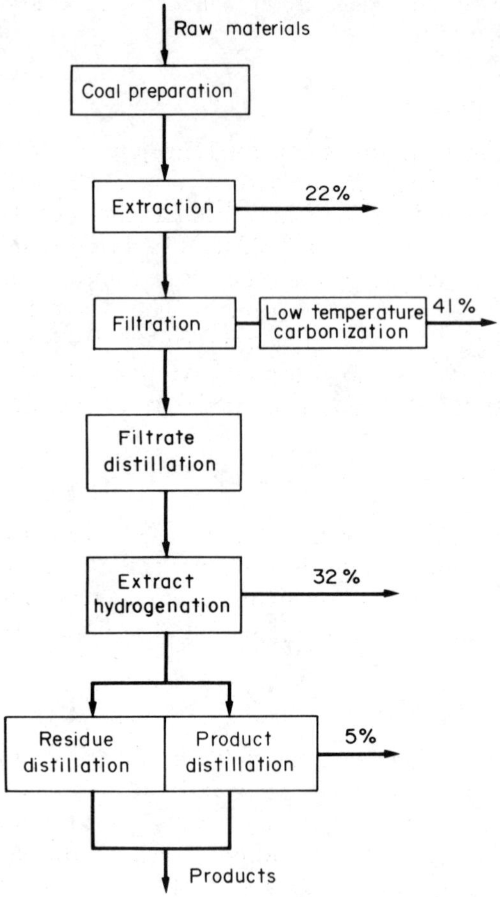

FIGURE 4.5 *Percentage discharge of effluents produced during a coal extraction liquefaction process*

weight. The relative magnitude of different sources is shown in Figure 4.5. Analyses for the different streams have been published (Chmielowski and Labuzek, 1979).

Very large quantities of solid wastes are generated in a full-scale liquefaction plant. Up to thousands of tonnes per day of ash will be deposited from a 6000 t day^{-1} unit using coal with an ash content of 5–40%. A considerable amount of spent catalyst also has to be taken care of, especially if the process uses a once-through catalyst. The H-coal process has a catalyst consumption of 1 t 200 t^{-1} of coal, for example, even though catalyst recycling takes place. Solid wastes will also be generated from the raw and waste-water treatment.

4 Control Aspects

Coal conversion techniques do not represent a very large industry today. However, large plants, converting 8–10 × 10^6 t yr^{-1}, are planned and in the U.S.A. it may be that the industry will involve 500 000 people by the year 2000 (DOE/HEW/EPA, 1979). The methods to be used are already developed and, from the environmental and health protection point of view, it is desirable that the emission control aspects are anticipated. To some extent new protection procedures and routines have to be developed and much can be gained from experience in the petrochemical and nuclear industries.

From a technical point of view, almost any environmental or health problem arising in the conversion industry can be solved. But a great deal depends on how much the industry can afford to pay, and the regulations and standards governments decide to enforce.

There are a number of possible control techniques that can be applied to potential gaseous, liquid and solid emissions.

4.1 Gaseous Emissions

Gasification and liquefaction of coal is able to reduce or eliminate unwanted emissions like nitrogen and sulphur. Conditions in the thermochemical process are highly reducing and sulphur is converted to substances like H_2S, COS and mercaptans, and nitrogen to NH_3. These gases are produced at a high pressure and may be easily removed. This is often necessary also for subsequent methanol or methane synthesis as the sulphur content in methanol synthesis gas has to be less than 0.1 ppm to avoid catalyst inactivation. The following account deals mainly with sulphur but there are effective methods for control of other substances also. For example, activated carbon can be used to remove most hydrocarbon fractions.

4.1.1 In-process sulphur removal and control

Sulphur in the raw product gas from a gasification or a hydrogenation step can be removed almost completely in two stages. The first (Table 4.6) brings down the sulphur content to a few ppm and the second (zinc oxide) further reduces the content below 0.1 ppm.

TABLE 4.6 *Processes for sulphur removal from raw or synthesis gas streams (Cavanaugh, Corbett and Page, 1977; Gas Processing Handbook, 1979)*

Alkazid	MEA
Amisol	MDEA
Benfield	Purisol
DGA	Rectisol
Estasolvan	Selexol
Fluor solvent	SNPA-DEA
Giammarco-Vetrocoke	Sulfinol

The first step yields a sour gas which has to be cleaned by use of a Claus or a Stretford unit, producing elemental sulphur and a tail gas. The tail gas from the Stretford unit is fairly clean and contains less than 8 ppm H_2S and less than 75 ppm of COS. The tail gas from the Claus unit, however, needs more treatment and one of the processes shown in Table 4.7 can be used to further reduce the sulphur emission. In total the sulphur leaves the conversion process as elemental sulphur, zinc sulphide and small amounts of H_2S and SO_2 (Cavanaugh, Corbett and Page, 1977; Gas Processing Handbook, 1979).

TABLE 4.7 *Processes for cleaning tail gas from a Claus unit (Cavanaugh, Corbett and Page, 1977; Gas Processing Handbook, 1979)*

Ammonium bisulphite/ ammonium thiosulphite	Gives less than 900 ppm SO_2 in off-gas
Beavon	Gives less than 10 ppm H_2S in off-gas
BSR/Selectox	Gives sulphur removal of 98-99% when used with a Claus unit
Clear Air	Gives less than 50 ppm SO_2 in off-gas
IFP Clauspol 1500	Gives less than 1500 ppm SO_2 in off-gas
SCOT	Gives less than 500 ppm H_2S in off-gas
Sulfreen	Gives sulphur removal of 99% when used with a Claus unit
W-L SO_2 recovery	Gives less than 100 ppm SO_2 in off-gas

4.1.2 Flue gas control

Flue gases contain sulphur and nitrogen, as oxides, which can be removed by scrubbing with alkaline solutions or by contacting with solid calcium carbonate, dolomite or manganese dioxide, as is practised for power or heat generation from coal or oil (Hoffman, 1978).

4.2 Liquid Emissions

The primary liquid emissions from coal conversion processes are often heavily contaminated by ammonia, sulphides and other inorganics, as well as by phenolic and various other organic compounds. The chemical oxygen demand (COD) and the biochemical oxygen demand (BOD) are often very high. Stutsman (1976) reports that the effluents from the Synthane gasification process and from the SRC liquefaction process can have BOD values as high as 30 000 mg l^{-1} and COD values up to 30–40 000 mg l^{-1}. The heavy metal content is usually relatively low.

Waste-water cleaning may have to be done in several stages: separation of particulate and emulsion phases; neutralization; removal of dissolved inorganics; removal of dissolved organics by adsorption or extraction; removal of organics by biochemical oxidation; a last step of ion exchange methods to remove inorganic salts.

4.2.1 Phase separation (Cavanaugh, Corbett and Page, 1977)

Flocculation–Flotation

By this widely used method the BOD and COD may be reduced by 80%, suspended solids by 75% and free oils by 97%.

Oil–water separators

The use of this simple technique, utilizing the difference in density between the oil and the water, can reduce the oil content by 60–99% and the solids by 10–50%.

Filtration

When sand is used as the filter medium the reduction may be 36% for BOD, 25–44% for COD, 52–83% for suspended oils and 70–75% for suspended solids.

4.2.2 Neutralization

The use of inorganic acids for neutralization of the alkaline waste-waters is effective but causes problems with corrosion. Saturation with CO_2 has been shown to be a more advantageous alternative (Chmielowski and Labuzek, 1979).

4.2.3 Removal of dissolved inorganics (Cavanaugh, Corbett and Page, 1977)

Acid gas stripping can be used to remove acid gases like CO_2 and H_2S, as well as some other substances, from the water. A removal efficiency of 99% for H_2S, 90% for NH_3 and 20–90% for phenol is reported. A special version (WWT) can be used to reduce the concentration of NH_3 below 50 ppm and of H_2S below 10 ppm.

4.2.4 Removal of dissolved organics by extraction or adsorption (Cavanaugh, Corbett and Page, 1977)

Extraction–Phenosolvan

This widely used technology is based on liquid–liquid extraction with butyl acetate, isopropyl ether or a light aromatic oil. The efficiency of removing phenols is high. The following figures are given: removal of monohydric phenols 99.5%; polyhydric phenols 60%; other organics 15%. The method is used at the SASOL plant in South Africa and an efficiency of 95% for total organics removal is reported.

Adsorption

Activated coal or synthetic polymers can be used as adsorbents. Removal efficiency for phenol is 99% and for COD, 81%.

TABLE 4.8 *Removal efficiencies (percentages) of the activated sludge and aerated lagoon biological oxidation processes (Cavanaugh, Corbett and Page, 1977)*

	Sulphides	BOD	COD	Phenols
Activated sludge	97–100	88–90	60–85	95–99
Aerated lagoon	95–100	75–95	60–85	90–99

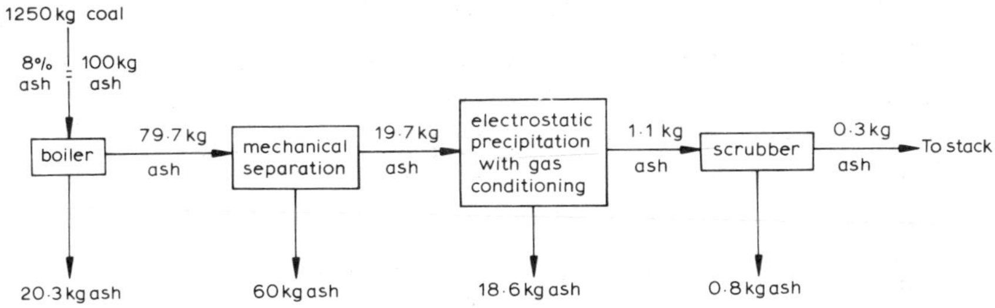

FIGURE 4.6 *Measured values for particulate removal systems for a large coal-fired power plant (Hoffman, 1978)*

4.2.5 Removal of organics by biological oxidation

This microbial treatment is sensitive to pH levels in the feed. A proper pretreatment of the effluent is important (EPA, 1975). It should be ensured that the concentration of NH_3, H_2S and extractable substances is as low as possible in order to avoid inactivation of the bacteria. A number of processes exist and the removal efficiencies for two of them are given in Table 4.8.

A number of substances derived from coal processing (cresols) are, however, characterized by a considerable resistance to the action of micro-organisms.

4.3 Solid Emissions

The solid wastes from coal conversion processes are derived from coal ash, spent catalyst and solid sludges. Only part of the ash is emitted to the atmosphere when coal is used as fuel for steam or heat generation. Figure 4.6 shows the effectiveness of three different particulate removal systems. The problem is rather to do with the solid waste when it is collected, not so much how to collect it. Spent catalysts can sometimes be recovered because of the high content of these valuable metals.

Sixty-five million tonnes of coal ash per year are produced by electric utilities in the U.S.A. (OECD, 1980). This ash is disposed of either in ponds or as landfill. Fly ash can be mixed with cement. This can be, however, only a partial solution to disposal as in the U.S.A., for example, cement production is 80×10^6 t yr^{-1} and the fly ash production is 45×10^6 t. The maximum ash : clinker ratio is 1 : 5.

5 References

ANWER, J. and BÖGNER, F. (1976). Kohlevergasung im Fluidatbett unter Druck. *Brennstoff-Wärme-Kraft,* **28,** 57–60.

ARNDT, E., FISCHER, R., FRÖHLICH, W., JÜNTGEN, H., TEGGERS, H. and WEISBRODT, I. (1979). Synthetisches Erdgas aus Kohle und Hochtemperatur-Wärme. *Erdöl & Kohle, Erdgas, Petrochemie,* **32,** 17–23.

BANICHEK, I.N. (1973). The Winkler process for production of low Btu gas from coal. *Clean Fuels from Coal,* Symposium (IGT), Chicago, Ill.

BATCHELOR, J.D. (1979). *Technical Status of the Large Coal Liquefaction Programs at the Department of Energy.* Unpublished Report, The Department of Energy, Pittsburgh, Pa.

BATTELLE (COLUMBUS LABORATORIES) (1973). *Detailed Environmental Analysis Concerning a Proposed Coal Gasification Plant for Trans-western Coal Gasification Company, Pacific Coal Gasification Company, Western Coal Gasification Company.* Before the Federal Power Commission, Washington, DC.

BECKER, P. (1979). Lösung der Umweltprobleme bei Grossanlagen zur Druckvergasung von Kohle nach dem Lurgi Verfahren. Vortrag gehalten anlasslich der 4. Sitzung des Fachausschusses *Vergasung, Verglussingung and Neue Produkte in Essen.*

BECKNER, J.L. (1975). Trace element composition and disposal of gasifier ash. *Proceedings of Seventh Synthetic Pipeline Gas Symposium,* Chicago, pp. 359–67.

BODLE, W.W. and VYAS, K.C. (1975). Clean fuels from coal – Introduction to modern processing. *Clean Fuels from Coal,* Symposium (IGT), Chicago, Ill.

BUDDEN, K.T. and ZIEGER, W.H. (1978). *Environmental Assessment of Coal Liquefaction,* Annual Report, EPA-600/7-78-019.

CAVANAUGH, E.C., CORBETT, W.E. and PAGE, G.C. (1977). *Environmental Assessment Data Base for Low/Medium/BTU Gasification Technology,* vol. II. Appendices A-F, USEPA-600/7-77-125b.

CHMIELOWSKI, J. and LABUZEK, S. (1979). The biochemical treatment of waste waters from the coal liquefaction process, *UN/ECE Symposium on the Gasification and Liquefaction of Coal,* Katowice.

CONSIDINE, D.M. (1977). *Energy Technology Handbook.* McGraw-Hill, U.S.A.

COOKE, R. and GRAHAM, P.W. (1965). The biological purification of the effluent from a Lurgi plant gasifying bituminous coals. *International Journal of Air and Water Pollution,* **9,** 97–112.

DAVY POWERGAS SALES BROCHURE (1973), *Winkler Generator Units,* 21e/6/73.

DE VAUX, G.R. and DUTKIEWICZ, B. (1979). Status of the H-coal process, *UN/ECE Symposium on the Gasification and Liquefaction of Coal,* Katowice.

DOE/HEW/EPA (1978). *Health and Environmental Effects of Coal Gasification and Liquefaction Technologies.* Background material for a workshop (ed. by J. Antizzio). DOE/HEW/EPA-01-M78-58.

DOE/HEW/EPA (1979). *Health and Environmental Effects of Coal Gasification and Liquefaction Technologies,* A workshop summary and panel reports (ed. by R. Brown and A. Witter). DOE/HEW/EPA-03-MTR-79W00137.

DOLCH, P. (1936). *Wassergas, Chemie und Technik.* J.A. Barth Verlag.

ELGIN, D.C. (1976). Conventional gasification technologies, *ECE Symposium on Gasification and Liquefaction of Coal,* Dusseldorf.

EPA (1975). Evaluation of pollution control in fossil fuel conversion processes, Liquefaction, Section 2. SRC Process. U.S. EPA Office of Research and

Development, Washington, DC, pp. 39.

EPPERLY, W.R. and TAUNTAN, J.W. (1979). *Exxon Donor Solvent Coal Liquefaction Process Development.* American Chemical Society Conference, Colorado Springs.

FALBE, J. (1977). *Hrg Chemierohstoffe aus Kohle.* Georg Thième Verlag, Stuttgart.

FEISTEL, P.P., VAN HEEK, K.H. and JÜNTGEN, H. (1978). *German Chem. Eng.*, **5**, 294–8.

FLESCH, W. and VELLING, G. (1962). Die Vergasung von Kohle im Winkler-Generator. *Erdöl und Kohle, Petrochemie,* **15** (9), 710–13.

FORNEY, A.J., HAYNES, W.P., GASIOR, S.J., JOHNSON, G.E. and STRAKEY, J.P. JR (1974). *Analysis of Tars, Chars Gases and Water Found in Effluents from the Synthane Process.* U.S. Bureau of Mines, Technical Progress Report TPR 76.

FRANKE, F.H. (1978). Vergasung von braunkohle zu synthese und reduktionsgas. *Vortragsveröffentlichungen, Haus der Technik,* **H. 405**, 60–8.

FRANZEN, J.E. (1977). *Einige Uberlegungen zur Kohlevergasung.* Tech. Mitt, Krupp-Forsch.-Ber., pp. 83–94.

FREDERSDORFF, VON, G.G., ELLIOTT, M.A. and LOWRY, H.H. (1963). *Chemistry of Coal Utilization.* Wiley, London.

FRIEDRICH, F., ROMEY, I. and STROBEL, B. (1980). Germany's new coal hydrogenation process – now proven by three years of Ruhr experimental plant experience. 88th AICHE Meeting, Philadelphia.

GAS PROCESSING HANDBOOK (1979). *Hydrocarbon Processing,* April 1979, pp. 99–170.

GHASSEMI, M., CRAWFORD, K. and QUINLIVAN, S. (1978). *Environmental Assessment Data Base for High-Btu Gasification Technology,* vol. II, Appendices A, B, and C. U.S. Environmental Protection Agency Report EPA-600/7-78-186b, Washington, DC.

GHASSEMI, M., CRAWFORD, K. and QUINLIVAN, S. (1979). *Environmental Assessment Report; Lurgi Coal Gasification Systems for SNG.* U.S. Environmental Protection Agency Report EPA-60037-79-120, Washington, DC.

GRATKOWSKI, VON, H.W. (1958). Kohlvergasungsverfahren. *Ullmans Encyklopadie der Techn Chemie,* **10.** Urban & Schwarzberg, Munchen, Berlin, pp. 376–458.

GÖKE, E. (1979). Technische Entwicklung der Kohle-Flugstrom-Vergasung und ihre wirtschartlichen Perspektiven. Vortrag: *United Nations Symposium on World Coal Prospects,* Katowice.

HALL, H.J. (1974). Trace elements and potential toxic effects in fossil fuels. *EPA Symposium, Environmental Aspects of Fuel Conversion Technology,* St Louis.

HEDDEN, K. (1958). Kohlevergasung, Phys.-chem. Grundlagen. *Ullmans Encyklopadie der techn. Chemie,* **10.** Urban & Schwarzberg, Munchen, Berlin, pp. 362–76.

HEDDEN, K. (1962). Methanbildung aus Wasserstoff und Kohlenstoff bei hohen Temperaturen und Druken. *Zeitschrift fur Elektrochemie, Berichte der Bundengesellschaft fur Physikalische Chemie,* **66,** 652–61.

HEDDEN, K. and MIENKINA, G. (1965). Zur Reaktionskinetik der Druckvergasung von Kohlenstoff mit H_2O–H_2-Gemischen. *Brennstoff-Chemie,* **46,** 366–71.

HEEK, VAN, K.H. (1978). *Stand der Kohlevergasung.* Vortragsveröfflentlichungen Haus der Technik, **H.405,** 50–9.

HEEK, VAN, K.H., JÜNTGEN, H. and PETERS, W. (1973). Fundamental studies on coal gasification in the utilization of thermal energy from nuclear high-temperature reactors. *Journal of the Institute of Fuel,* **249,** 249–58.

HILLER, H. (1975). Kohle-Druckvergasung. *Erdöl und Kohle,* **28,** 74–80.

HOFFMAN, E.J. (1978) *Coal Conversion.* Modern Printing Company, Laramie, Wy.

HOOGENDORN, J.C. (1976). Die neue Anlage Zur Erzeugung glussiger Kohlenwasserstoffe aus Kohle in Sudafrika. *Gas Warme Internat.,* **25,** 283.

HULISZ, ST, PAWLAK, ST, STAWIANY, W. and SWIADROWSKI, J. (1979). Hydrogenation of coal extracts and low-temperature tars in experimental installations in Poland. *UN/ECE Symposium on the Gasification and Liquefaction of Coal,* Katowice.

IHNATOWICZ, M. (1979). Investigations on the hydrogenation of coal extracts and low-temperature tars with the use of various catalysts. *UN/ECE Symposium on the Gasification and Liquefaction of Coal,* Katowice.

JAHNIG, C.E. (1975a). Evaluation of pollution control in fossil fuel conversion processes. Gasification, Section 5: BIGAS process, Report no. EPA 650/2-74-009g, Washington, DC.

JAHNIG, C.E. (1975b). Evaluation of pollution control in fossil fuel conversion. Gasification, Section 6: HYGAS process, Report no. EPA 650/2-74-009h, Washington, DC.

JÜNGTEN, H. (1977). Entwicklungstendenzen der Kohlevergasung. *Bergbau,* **28,** 347–55.

JÜNTGEN, H. (1979a). Theoretische Grundlagen von Vergasungsprozessen. *VGB-Kraftwerkstechnkik,* **59,** 557–64.

JÜNTGEN, H. (1979b). Verwendung von Kohlengas und Kohlenöl. *Gluckauf,* **115,** 329–38.

JÜNTGEN, H. and KLEIN, J. (1976). Purification of wastewater from coking and coal gasification plants using activated carbon. *Energy Sources,* **2,** 311–30.

JÜNTGEN, H. and VAN HEEK, K.H. (1977). A technical scale gas generator for steam gasification of coal using nuclear heat. *Nuclear Technology,* **35,** 581–90.

JÜNTGEN, H. and VAN HEEK, K.H. (1979a). Allothermal pressurized gasification in a fluidized bed. *Informationstagung: Neue Verfahren zur Kohleveredlung.* Luxemburg.

JÜNTGEN, H. and VAN HEEK, K.H. (1979b). An update of German non-isothermal coal pyrolysis work. *Fuel Processing Technology,* **2,** 261–93.

JÜNTGEN, H., KLEIN, J., KNOBLAUCH, K., SCHROTER, H.J. and SCHULTZE, J. (1980). Other uses of coal. *Chemistry of Coal Utilization* (ed. by H.H. Lowry).

KETCHAM, N.H. and NORTON, R.W. (1960). The hazards to health in the hydrogenation of coal. III: The industrial hygiene studies. *Archives of Environmental Health,* **1,** 194–207.

KOENIG, J.C. and SCHMALZER, D.K. (1979). An overview of SRC process Development and Commercial Scale Development of the SRC II Process. *UN/ECE Symposium on the Gasification and Liquefaction of Coal,* Katowice.

KRAAIJVELD, H.J. (1978). *Technische und wirtschaftliche Aspekte des Shell/Koppers-Kohlevergasungsverfahrens.* Vortragsveroffentlichungen, Haus der Technik, **H.405,** pp. 74–7.

KRICZKO, A.A. (1977). Problemy kompleksnogo ispolzowania ugla. *Chimja Twiordogo Topliwa,* **6,** 3–10.

KRICZKO, A.A. and JULIN, M.K. (1973). Chimiczeskije i Topliwnyje producty iz smiesiej ugla i niefty. *Chimja i Pierorabotka Topliw,* **29,** 29–43.

LANGHOFF, J., WOLOWSKI, E., ESCHER, G. and HOSANG, H. (1979). Further development of German technology coal hydrogenation bÿ Ruhrkohle AB and VEBA – Chemie AG – The 200 t/d pilot plantBottrop. *UN/ECE Symposium on the Gasification and Liquefaction of Coal,* Katowice.

LINKE, A. and VOGT, E.V. (1979). *Die Druckvergasung von Kohle im Flugstrom nach Shell/Koppers.* Vortrag: Achema, Frankfurt.

MAGEE, E.M., JAHNIG, C.E. and SHAW, H. (1974). Evaluation of pollution control in fossil conversion processes. Gasification, Section 1; Koppers-Totzek Process. Report no. EPA 650/2-74-009a, Washington, DC.

MASSEY, M.J. and DUNLOP, R.W. (1976). Environmental assessment in coal gasification processing. *Proceedings of Eighth Synthetic Pipeline Gas Symposium,* Chicago, pp. 205–19.

MOELLER, F.W., ROBERTS, H. and BRITZ, B. (1974). Methanation of coal gas for SNG. *Hydrocarbon Processing,* **53**, 69–74.

MOSCHITTO, R.D. (1978). Operation of the Fort Lewis, Washington SRC pilot plant in the SRC I and SRC II processing modes. *13th Intersociety Energy Conversion Engineering Conference,* San Diego, California.

NAS (1977). Assessment of technology for the liquefaction of coal, National Academy of Sciences, U.S.A., December 1977.

NIOSH (National Institute for Occupational Safety and Health) (1978a). *Recommended Health and Safety Guidelines from Coal Gasification Pilot Plants.* NIOSH Publication, pp. 78–120. Cincinatti, Ohio.

NIOSH (National Institute for Occupational Safety and Health) (1978b). *Criteria for a Recommended Standard: Occupational Exposures in Coal Gasification Plants.* NIOSH Publication, pp. 78–191. Cincinatti, Ohio.

NOWACKI, P. (1979). *Coal Liquefaction Processes.* Noyes Data Corporation, Park Ridge, New Jersey.

OECD (1980). Coal and the environmental advanced control technologies. Group on Energy and Environment. OECD-ENV/EN/80.22, Paris.

PARKER, C.L. and DYKSTRA, D.I. (1978). *Environmental Assessment Data Base for Coal Liquefaction Technology,* vol. II. Synthoil, H-Coal, Exxon Donor Solvent Processes. U.S. Environmental Protection Agency Report EPA 600/7-78-184b, Washington, DC.

PETERS, W. (1976). Grundlagen der Kohlevergasung. General Report. Symposium uber Kohlevergasung und -verflussigung der UN-Economic Commission for Europe, Dusseldorf.

PETERS, W. (1979). Die Kohleverflussigung in der Bundesrepublik Deutschland. *Gluckauf,* **115** (8), 325–9.

PETERS, W. and SCHILLING, H.D. (1977). Perspektiven der Kohlevergasung under der Kohlevergasung aus der Sicht der Steinkohle. *Erdol & Kohle,* **30**, 552–7.

ROBSON, B. (1977). A review of gasification of power generation. *Energy Research,* **1**, 157.

RÖBKE, G. (1978). *Weiterentwicklung des Lurgi-Druckvergasungsverfahrens.* Vortragsveröffentlichungen, Haus der Technik, **H.405**, 85–90.

ROGERS, S.E., MARROCCO, N.J., AKHTAR, S. and JAVORSKY, P.M. (1978). Coal liquefaction in synthoil reactor without added catalyst, 85th National AIChE Meeting, Philadelphia. Pa.

ROLKE, D. (1979). Lösung der Umweltfragen in Lurgi-Kohledruckl Vergasungs-Anlage. Vortag gehalten anlasslich der 4. Sitzung des Fachausschusses. *Vergasung, Verflussigung und Neue Produkte in Essen.*

ROSSBACH, M., KUFFNER, P., FUSSMAN, G., GRAMS, W. and PETER, K. (1978). *Weiterentwicklung des Lurgi-Druckvergasungsverfahrens.* Vortragsveröffentlichungen, Haus der Technik, **H.405**, 78–84.

SCHILLING, H.D., BONN, B. and KRAUSS, U. (1979). *Kohlevergasung.* Band 4, Rohstoffindustries International. Verlag Gluckauf GmbH, Essen.

ROLKE, D. (1979). Lösung der Umweltfragen in Lurgi-Kohledruckl Vergasungs-Anlage. Vortag gehalten anlasslich der 4. Sitzung des Fachausschusses. *Vergasung, Verflussigung und Neue Produkte in Essen.*

ROSSBACH, M., KUFFNER, P., FUSSMAN, G., GRAMS, W. and PETER, K. (1978). *Weiterentwicklung des Lurgi-Druckvergasungsverfahrens.* Vortragsveröffentlichungen, Haus der Technik, **H.405,** 78–84.

SCHILLING, H.D., BONN, B. and KRAUSS, U. (1979). *Kohlevergasung.* Band 4, Rohstoffindustries International. Verlag Gluckauf GmbH, Essen.

SCHMIDT, J. (1966). *Technologie der Gaserzeugung.* Band II, Vergasung: VEB Deutscher Verlag fur Grundstoffindustrie, Leipzig.

SCHMIDT, C.E., SHARKEY, A.G., JR. and FRIEDEL, R.A. (1974). Mass spectrometric analysis of product water from coal gasification. U.S. Bureau of Mines Technical Progress Report TPR 86.

SCHULZE, J. (1974). Qualitative und quantitative Substitutionschancen fur die Kohlechemie. *Chem-Ing.-Techn.,* **46,** 976–80.

SCHULZE, J. (1978). *Die Erzeugung von Chemierohstoffen aus Kohle.* Vortragsveröffentlichungen, Haus der Technik, **H.405,** 16–29.

SEIPENBUSCH, J. and RUPRECHT, P. (1978). *Das RAG/RCH-Projekt Kohlestaub-Druckvergasung,* Vortragsveroffentlichungen, Haus der Technik, **H.45,** 69–73.

SEXTON, R.J. (1960a). The hazards to health in the hydrogenation of coal. I. An introductory statement on general information, process description, and a definition of the problem. *Archives of Environmental Health,* **1,** 181–6.

SEXTON, R.J. (1960b). The hazards to health in the hydrogenation of coal. IV. The control program and the clinical effects. *Archives of Environmental Health,* **1,** 208–31.

SHAW, H. and MAGEE, E.M. (1974). Evaluation of pollution control in fossil fuel conversion processes, Gasification, Section 3: Lurgi Process, Report no. EPA 650/2-74-009c, Washington, DC.

STAEGE, H. (1976). Die Kohlevergasung nach Koppers-Totzek. *Erdöl-och Erdgas-Zeitschrift,* **92,** 381–7.

STOTLER, H.H., MACARTHUR, J.B. and COMOLLI, A.G. (1980). Current status of H-coal commercialization. *Seventh International Conference on Coal Gasification, Liquefaction and Conversion to Electricity,* Pittsburg, Pa.

STROBEL, B., ROMEY, I. and KOELLING, G. (1979). German routes to coal hydrogenation. *Advances in Coal Utilization Technology.* Louisville, Kentucky.

STUTSMAN, M.J. (1976). *Water Conservation and Pollution Control in Coal Conversion Processes.* Water Purification Associates, Cambridge, Mass.

TOPPER, J.M. (1980). UK coal liquefaction. CRE, Cheltenham. *Energy Digest,* June 1980.

URQUHART, D.B., MARTIN, T.G. and WHITEHEAD, J.C. (1979). Preparation of refined liquid products from coal by the supercritical gas extraction route, *UN/ECE Symposium on the Gasification and Liquefaction of Coal,* Katowice.

VÖLKEL, H.K., ECKSTEIN, G., LINKE, A. and VOGT, E.V. (1979). Operation and economics of the Shell-Koppers coal gasification process. *14th World Gas Conference of the International Gas Union,* Toronto.

VGB KONFERENZ (1979). Kohlevergasung in der Energietechnik, VGB-TB 201, Essen.

WADDEN, R.A. (1976). Coal hydrogenation and environmental health. *Environmental Health Perspectives,* **14,** 201–6.

WOLK, R.W. (1979). Overview of Liquefaction Process Technology. Unpublished Report, The Department of Energy, Pittsburgh, Pa.

WÜRFEL, H. and JORZYK, S. (1979). The Saarberg coal liquefaction process, *UN/ECE Symposium on the Gasification and Liquefaction of Coal,* Katowice.

YOON, H., WEI, J. and DENN, M.M. (1978). A model for moving bed coal gasification reactors. *AIChE Journal,* **25,** 885.

5

Application of the Exposure Commitment Model to Coal Utilization Emission Problems

1 The Exposure Commitment Concept

In order to assess the total effects of releases of radioactive material into the environment by such practices as detonating atomic devices, the dose commitment concept was developed by Lindell (1960) and the United Nations Scientific Committee on the Effects of Atomic

Radiation (UNSCEAR). Information on the sources and effects of ionizing radiation are published from time to time (UNSCEAR, 1977). It was subsequently suggested by Butler *et al.* (1972), and Lindell (1978) that the concept could be applied to non-radioactive environmental pollutants. The concept has been developed into a method, now called the exposure commitment method (Barry, 1979; O'Brien, 1979; Bennett, 1981; Buchanan and Bennett, 1980), to give a time-independent approach to environmental pollutant assessment. A brief outline of the concepts and definitions of the method are given here, followed by a discussion of its applicability to coal combustion and conversion activities.

1.1 Concepts and Definitions

Two recent publications on the exposure commitment method (Bennett, 1981; Buchanan and Bennett, 1980) emphasize that exposure commitment is a combined measure of the concentration and duration of a pollutant's presence in an environmental medium. In the entire period of a particular exposure:

$$E = \int_0^\infty C(t)\mathrm{d}t$$

where $C(t)$ is the concentration at any particular time t. If the probability of occurrence of a response rather than the severity, increases with exposure (as is postulated with radionuclide exposure–response relationships) then the curve relating concentration and time (CT) is one of equal probability effects. This is referred to as a stochastic effect. If a "threshold" is postulated, but stochastic effects are exhibited once this is exceeded, the concentration × time curve will indicate threshold exposure. For either, the exposure commitment is time-independent: various combinations of concentration and time periods can give the same exposure commitment and equal probability of effect or transcendence of the threshold.

Exposure commitment may be applied for the site of impact but it can also be used as a measure of the exposure in the preceding medium. Thus intake may be related to the external conditions and also transfer from one environmental compartment to another quantitatively described.

If linearity is assumed for the transfer process, transfers are additive and intercompartmental movement (from compartment i to compartment j) of the pollutant may be described by its flux:

$$F_{ij}(t) = K_{ij}Q_i(t)$$

where $Q_i(t)$ is the amount of pollutant in the donor compartment and K_{ij} the proportionality constant (the transfer rate constant), depending upon the ratio of concentrations $C_i(t)$ and $C_j(t)$. Then:

$$F_{ij}(t) = K_{ij}M_i\frac{Q_i(t)}{M_i} = K_{ij}M_iC_i(t)$$

where M_i is the mass or volume of compartment i.

The intake to a compartment is the integral of the flux (the product of the average amount passing per unit time and the time period). Hence the total intake can be expressed as:

$$\begin{aligned}\Sigma_i I_{ij} &= \Sigma_i \int_0^\infty F_{ij}(t)\mathrm{d}t \\ &= \Sigma_i K_{ij} M_i \int_0^\infty C_i(t)\mathrm{d}t \\ &= \Sigma_i K_{ij} M_i E_i\end{aligned}$$

The total intake is directly proportional to the exposure commitment in the donor compartment and is thus a useful measure of environmental transfer. As intake commitment and exposure commitment are integral quantities, time integrals of flux in concentration remove time-dependence and give an expression of the overall transfer results.

Various integral quantities can be related by means of a transfer factor (P_{ij}). For example:

$$P_{ij} = \frac{E_i}{E_j} = \frac{\int_0^\infty C_j(t)\mathrm{d}t}{\int_0^\infty C_i(t)\mathrm{d}t}$$

For continuing releases steady-state levels in compartments are commonly attained. With concentrations that remain fairly constant, transfer factors may be evaluated by taking the ratios of equilibrium concentrations on fluxes.

1.2 *Evaluation of the applicability of the methods*

In terms of measuring or assessing likely effects, the exposure commitment method must be applied with caution to coal combustion and conversion impacts. Various combinations of concentration and time period will give an equal exposure commitment, but these may not be associated with equal occurrences of harmful effects. This will be particularly true if threshold levels apply. There is a considerable body of opinion that this is so for most emissions from coal combustion and conversion (Ulmer, 1977) and that no linear dose–response relationship exists. Repair mechanisms following exposure are cited as an added reason for rejecting the "no zero-effect dose" hypothesis.

However, others have argued in favour of the linear no-threshold model (LNTM), as a working hypothesis, on the basis of synergistic effects, the carcinogenic effect of some products of coal combustion and conversion, and the wide confidence limits that need to be set on data (Butler, 1978; Intl. Symp., 1978; Wilson, 1978). Some have concluded that for chemical pollutants of air it would be prudent to assume a linear no-threshold model.

As explained, this assumption permits the application of the exposure commitment method with relative ease. It is also worth noting that the LNTM has another important consequence for pollution control practices. If the quantity of a pollutant to which a population is exposed is constant, the total number of health effects produced is constant (with an uncertainty of the square root of the number of effects). Thus no benefit results from measures that expose larger numbers of people to lower concentrations as results from using higher stacks for air effluents and discharging liquid effluents into larger volumes of water. The only control measures that succeed are those that prevent pollutants from coming in contact with receptors such as retention of the pollutant at the source or, for polluted air, discharging it at the seaboard where it will be carried out to the ocean. But even this may have its limitations. Another biological assumption basic to the exposure commitment method, already referred to, is that there is no repair of the effects, however small. This means that the effects per unit exposure will be the same, whether this is over a short or a prolonged period.

Following on these two assumptions it is possible to conclude that all exposure commitments to a population, however large, over all time, however long, contribute to the total impact. Thus for every action that contaminates the environment the total CT for the receptor must be assessed together with the total of the resulting effects or harm, both short-term and long-term. The various types of situations to be considered have been classified as follows (Butler, 1978):

Release	*Contaminant*
A. Single	1. Short-lived 2. Long-lived
B. Continued	1. Short-lived 2. Long-lived

For case A.1 the annual exposure commitment will encompass the total and will be adequate to assess the total detriment. For the other three cases the estimates must extend beyond 1 year. The assessment of effects beyond 1 year may be necessary because:

(1) the amount of the practice may increase in the future;
(2) the pollutant may persist in the environment long after its release;
(3) the pollutant may be produced in the environment long after the release of its parent substance, e.g. methyl mercury;
(4) the pollutant may remain in the body of the receptor long after its intake;
(5) the biological effects may be expressed long after the initial intake.

The information required for calculating exposure commitment usually consists of data obtained by survey, varying in time and space, on sources, pollutants released, receptors and information from research on the organs and processes of the receptor affected by the pollutant. However, it requires a more limited data base than time-dependent assessments. Exposure commitment, multiplied by the appropriate risk estimate, allows an estimate of the harm commitment to be obtained.

The harm commitment method is most useful for assessment and planning purposes when each estimate is normalized to the amount of practice that generates the exposure commitment. When given in this way, the commitments from alternative practices can be compared and the consequences of future expansion can be calculated.

2 Application of the Exposure Commitment Model

Non-stochastic effects that resulted from the burning of fossil fuels were the increased deaths that arose promptly from the heavy fogs that occurred in London, Oslo and the Meuse valley (Hamilton, 1980). These were probably due to the combined action of soot with oxides of sulphur and nitrogen.

The stochastic effects of most concern in radiation protection are increased malignancies and mutations. The pollutants released to the atmosphere by combustion of fossil fuels, that may affect human health, are: oxides of carbon, nitrogen and sulphur; heavy metals; polynuclear aromatic hydrocarbons such as benzo(α)pyrene; radionuclides. Some of these are probably carcinogenic or mutagenic. Another kind of stochastic effect, probably caused by oxides of sulphur (sulphate) combined with soot and aggravated by cigarette smoking, is the development of chronic bronchitis after long exposure (Nishiwaki *et al.*, 1971).

TABLE 5.1 *Annual effluent production (tonnes) from a 1000 MW(e) coal-fired power plant (UNEP, 1981)*

Airborne effluents:	
Particulates	3×10^3
Sulphur dioxides	11×10^4
Nitrogen oxides	2.7×10^4
Carbon monoxide	2×10^3
Hydrocarbons	400
Liquid effluents:	
Organic material	66.2
Sulphuric acid	82.5
Chloride	26.3
Phosphate	41.7
Boron	331
Suspended solids	497
Solid wastes:	
Bottom ash + recovered fly ash	3.6×10^5

2.1 *Health effects of electric power generation*

Effluents from a coal-fired power plant generating 1000 MW(e) yr^{-1} (8.76×10^9 kWh yr^{-1}) are given in Table 5.1. They are calculated from data given by Wilson and Jones (1974) and assume the power plant burns 3×10^6 t coal with a 2% sulphur content, an energy content of 2.74×10^7 kJ t^{-1} and the thermal efficiency of the power plant is 38%; a fly ash removal efficiency of 99% and no flue gas desulphurization is assumed. Such a plant would have a capacity greater than 1000 MW(e), (a power plant with a generating capacity of 1000 MW(e), operating at 0.75 capacity would produce 6.75×10^9 kWh yr^{-1}).

Emissions from power plants in the Federal Republic of Germany, burning bituminous coal and conforming to the prevailing emission standards, show similar stack releases except that, due to some removal of sulphur and particulates in the flue gas, SO_2 and particulate values are considerably lower. Brewer (1976) estimates trace metal and radioactivity emissions from such a coal-fired power station as 0.67 t and about 1 GBq respectively. Table 5.2 gives comparative figures for different forms of fuel for airborne effluents. Emissions such as these contribute to levels of air pollution in the U.S.A. shown in Table 5.3.

The examples of power generation of burning coal, and the emissions that result, represent a long-term continuous process of type B (Butler,

TABLE 5.2 *Airborne effluents (tonnes) from 1000 MW(e) yr^{-1} power plants (UNEP, 1981)*

Fuel	Particulates	CO	NO_x	SO_2	Hydrocarbons
Coal	3×10^3	2×10^3	2.7×10^4	11×10^4	400
Oil	1.2×10^3	0.7×10^3	2.5×10^4	3.7×10^4	470
Natural gas	0.5×10^3	negligible	2×10^4	20.4	34

1978). The pollutants emitted, and their effects, are both short- and long-lived. A number of epidemiological studies have shown that an increase in airborne sulphates is associated with an increase in death rate; the results of several such studies are summarized by Hamilton (1980). There the risk estimate given is 33 excess deaths per year in a population of a million persons exposed to an increase in airborne sulphates of 1 μg m^{-3}.

To estimate the exposure commitment and harm commitment from these data it must be assumed that the emissions are constant and impinge on some population distributed around the power plant. For a population of 3 million uniformly distributed within a radius of 50 miles, Hamilton (1980) assumed an SO_2 emission rate of 0.12 lb (54 g) of SO_2 per million Btu in the fuel (< 1% sulphur coal with 90% removal of sulphur in flue gas). The calculated deaths per year from the generation of 1000 MW(e) (plant operating at 65% capacity) were: 0–3, mean = 0.6 in a 50-mile radius, and 0.30, mean = 6 for the whole U.S.A.

TABLE 5.3 *Concentrations of pollutants in urban and rural air (μg m^3) (Natusch, 1978)*

Substance	Urban air		Rural air
	Range	Average	Range
SO_2	20–1200	70	0.1–5
NO_x	50–400	100	2–6
Particulates	$(60–220) \times 10^3$	140×10^3	$(5–60) \times 10^3$

2.2 *Collective Dose Commitments Due to Radioactive Effluents from Power Stations*

2.2.1 Introduction

UNSCEAR (1977) have made a comparative radiation exposure assessment for coal and nuclear energy-based electricity generation. Revised calculations have recently been made (Halbritter *et al.*, 1981) for siting and technical conditions in the Federal Republic of Germany (F.R.G.). It is thus possible to use these calculations to compare a model coal-fired power plant and a model nuclear power plus full reprocessing and waste treatment plant. Comparisons of radiation exposure from the total nuclear fuel cycle and coal-fired power plants must include exposure due to accidents from nuclear sources, and exposure values for these are taken from a study undertaken in the F.R.G. (Ministry for Research and Technology, 1979).

2.2.2 Effluents and collective dose commitments for model plants

Calculations concerning radioactive effluents are standardized on a GW(e)yr basis for a bituminous coal-fired power plant and a nuclear fuel power plant together with all the relevant facilities. For the coal-fired power plant assumptions similar to those of McBride *et al.* (1977) were made and these lead to emission rates that agree well with those of UNSCEAR (1977). Measurements of fly ash activity (Chatterjee *et al.*, 1980; Kolb, 1979) are also of the same order of magnitude except that Pb-210 and Po-210 are highly enriched in fly ash where there is liquid ash removal, and such a condition was assumed for the model plant. Table 5.4 gives effluent values. Emission data for a typical PWR were based on actual emissions of three plants in the F.R.G. since 1972 (Table 5.5). For the fuel reprocessing and waste treatment plant a global estimation on the basis of release factors was made and is in good agreement with other estimates (USDOE, 1979). Effluent values are given in Tables 5.5 and 5.6.

To make a risk specific comparison collective dose commitments must be calculated. To allow comparison, values from UNSCEAR (1977) were converted to dose equivalents (ICRP, 1976). Effective dose commitments were obtained to take account of specific conditions in the F.R.G. (siting and technical features; population density, 250 km^{-2}) using dose conversion factors (ICRP, 1979).

TABLE 5.4 *Radionuclide emissions and collective dose equivalent commitment from coal-fired power plants*

Nuclide	Emissions (GBq per GW(e)yr)		Collective dose equivalent commitment (Man-Sv per GW(e)yr)			
	UNSCEAR (1977)	Halbritter *et al.* (1981)	UNSCEAR (1977)		Halbritter *et al.* (1981)	
			Ingestion	Inhalation	Ingestion	Inhalation
U–238	1.85	0.48	0.1		0.001	0.085
Th–234		0.27			–	–
U–234		0.48			0.002	0.17
Th–230		0.27			0.023	0.137
Ra–226	0.37	0.35	0.014		0.083	0.005
Pb–210	3.7	2.18	0.4		1.54	0.078
Po–210		5.55		no values given	0.91	0.098
Th–232	0.37	0.18	0.002		0.032	0.315
Ra–228	0.37	0.23	0.03		0.055	–
Th–228	0.37	0.18			0.005	0.105
Ra–224		0.23			0.003	0.003
K–40	5.55	0.78			–	–
TOTAL			0.546		2.65	0.996
					3.73[a]	

[a]Includes 0.08 γ soil radiation

TABLE 5.5 *Collective dose equivalent commitment due to radioactive emissions from nuclear power plants*

Nuclide	Exposure pathway	Radioactive emissions (GBq per GW(e)yr)		Collective dose equivalent commitment (Man-Sv per GW(e)yr)		
		UNSCEAR (1977)	Halbritter *et al.* (1981)	UNSCEAR (1977)	Halbritter *et al.* (1981)	
				Gonads	Gonads	Effective dose
Kr, Xe, Ar	atmosphere (local and regional)	3.7×10^7	3.5×10^5	2	0.09	0.09
H–3		6.7×10^5	1.4×10^3	0.04	0.0007	0.0007
C–14		2.2×10^2	2.2×10^2	0.006	0.015	0.015
Cs, Co, Sr, Ru		2.6×10^0	2.6×10^0	0.033	0.03	0.031
H–3	water (local and regional)	1.5×10^5	1.5×10^4	0.3	0.028	0.028
Cs, Co, Mn, I		2.2×10^2	8.1×10^1	0.06	0.032	0.033
H–3	atmosphere and water (global)	5.9×10^5	1.9×10^4	0.5	0.015	0.015
C–14		2.2×10^2	2.2×10^2	2.7	1.35	1.35
TOTAL				5.64	1.56	1.56

TABLE 5.6 *Collective dose equivalent commitment due to radioactive emissions from reprocessing plants*

Nuclide	Exposure pathway	Radioactive emissions (GBq per GW(e)yr)		Collective dose equivalent commitment		
		UNSCEAR (1977)	Halbritter *et al.* (1981)	UNSCEAR (1977)	Halbritter *et al.* (1981)	
				Gonads	Gonads	Effective dose
Kr–85	atmosphere (local and regional)	13.875×10^6	3.7×10^5	0.007	0.0005	0.0005
H–3		37×10^3	3.7×10^4	0.002	0.005	0.005
C–14		5.18×10^2	3.7×10^2	0.01	0.018	0.18
Cs, Ru, Sr		3.7×10^0	1.48×10^1	0.002	0.56	0.57
H–3	water (local and regional)	2.22×10^5	3.7×10^5	0.4	0.67	0.67
Cs, Ru, Sr, I		3.7×10^3	3.7×10^0	0.4	0.017	0.016
H–3	atmosphere and water (global)	2.59×10^5	4.07×10^5	0.22	0.34	0.34
Kr–85		13.875×10^6	3.7×10^5	0.9	0.025	0.025
C–14		5.13×10^2	3.7×10^2	6.3	2.25	2.25
TOTAL				8.24	3.89	3.90

TABLE 5.7 *Collective dose equivalent commitments and number of fatalities (somatic) for the nuclear fuel cycle and for coal-fired power plants*

Exposure	Collective dose equivalent commitment (Man-Sv per GW(e)yr)			Number of fatalities (Somatic) $(GW(e)yr)^{-1}$
	UNSCEAR (1977)	Halbritter *et al.* (1981)		
	Gonads	Gonads	Effective dose	
Nuclear power plant:				
Public	6	1.6[a]	1.6[a]	0.02
Occupational	10	4[a]	4	0.05
Reprocessing:				
Public	8	4	4	0.05
Occupational	12	1	1	0.01
Mining, milling, fuel fabrication and transportation:				
Public	0.2	0.2	5	0.06
Occupational	2	2	2	0.02
Total for nuclear fuel cycle (normal operation)	38.2	12.8	17.6	0.22
Nuclear accidents:				
Non-stochastic effects				4×10^{-5}
Stochastic effects			30	0.4
Coal-fired power plants	0.6	–	3.7	0.05

[a]for PWR plants

Generally the same model assumptions made by UNSCEAR (1977) were used, although for the coal-fired power plant some additional exposure pathways were taken into account.

It will be seen (Table 5.4) that the new estimates yield about 4 man-Sv per GW(e)yr as compared to about 0.6 man-Sv per GW(e)yr for coal-fired power plants. The higher exposure is due mainly to conservative assumptions in relation to the distribution of radionuclides in the soil that affect the ingestion pathway.

For the nuclear fuel cycle, modifications to suit conditions in the F.R.G. were made to the basic assumptions (UNSCEAR, 1977). Data from Kelly (1975) concerning C-14 were used, resulting in lower exposure values (Tables 5.5 and 5.6). The lower noble gas and tritium emissions expected from the PWR model power plant all contribute to lower exposure values than those given by UNSCEAR (1977).

2.2.3 Risk comparisons from radiation exposure from the use of coal and nuclear energy

Table 5.7 gives estimates of all exposures from different sectors of the nuclear fuel cycle. The 4 man-Sv per GW(e)yr occupational exposure is valid only for the PWR; other reactor types would yield 10 man-Sv per GW(e)yr, similar to the value given by UNSCEAR (1977). The radiation exposures made use of UNSCEAR (1977) data for mining, milling, fuel fabrication and transportation and extrapolated experience at Karlsruhe for fuel reprocessing. The new estimates show that the radiation exposure due to the emissions from coal-fired units is comparable to the non-occupational radiation exposure from all relevant stages of the nuclear fuel cycle at normal operation. Taking occupational and non-occupational exposures for normal operation, the contribution from the nuclear fuel cycle is about four times that of energy conversion by coal.

Potential accidents are a major nuclear energy risk. The nature of such accidents means that possible exposure values could be so different in relation to impact areas that both stochastic and non-stochastic health impacts must be considered, which is only possible using the detriment level approach. The F.R.G. risk study for nuclear power plants (Ministry for Research and Technology, 1979) gives stochastic risk (which is proportional to the expected dose commitment value) and is about 30 man-Sv per GW(e)yr (Table 5.7). Using dose-effect relationships for stochastic effects (ICRP, 1976) and estimates of non-stochastic accidental risk from the Ministry of

TABLE 5.8 *Effects of fuel cycle (per 1000 MW(e) plant-year, 65% capacity factor)*

Exposure	Coal effects: Deaths	Coal effects: Disease/injury	Nuclear effects: Deaths	Nuclear effects: Disease/injury
Mining				
Public	–	–	0.08	0.08
Workers:				
Accidents	0.6	42	0.31	12
Occupational disease	0.02–0.4	0.5–1	0.06 (cancer)	0.1–3
Processing				
Public	–	–	0.002	0.002
Workers:				
Accidents	0.05	2.9	0.004	1.3
Occupational disease	–	–	0.03 (cancer)	0.03 (cancer)
Transport				
Public	0.3–1.3 (Public and Workers: Accidents)	1.2–5.9 (Public and Workers: Accidents)	6×10^{-4}	6×10^{-4}
Workers:				
Accidents			0.1	0.1
Occupational disease	–	–	9×10^{-4}	9×10^{-4}
Electricity generation				
Public	0.2/million	NE	0.02	0.02
Workers:				
Accidents	0.1	3.3	0.01	1
Disease (cancer)	–	–	0.07	0.07
Catastrophic accidents	–	–	0.1	–
Waste management				
Public	–	–	5×10^{-5}	5×10^{-5}
Workers	–	–	7×10^{-4}	7×10^{-4}
Decommissioning	–	–	5×10^{-3}	0.07

NE = Not estimated

Research and Technology (1979) an estimate of the total health impact from radiation exposure from the two fuels can be made. This indicates that the risk from radiation caused by coal combustion is about an order of magnitude smaller than the total radiation exposure risk of nuclear power. There are also, however non-radiation risks associated with the use of the two fuels. These are dealt with briefly in the following sections.

2.3 *Comparative risks associated with total cycles*

The effects of electric power generation that have been considered are only part of the whole health impact of coal-fired electric power plants. It is necessary also to take into consideration the health effects of the complete operation with coal mining, processing, transporting and burning. These have been summarized for the U.S.A. by Hamilton (1980), as shown in Table 5.8, where they are compared with the same harm commitments from nuclear power generation (Hamilton, 1980). Tables 5.9 and 5.10 make a similar comparison for the U.S.S.R.

In making decisions about the control of a practice the scientific administrator will gain perspective by making other kinds of

TABLE 5.9 *Possible health damage resulting from a complete annual coal fuel cycle of a coal-fired power plant (per GW(e))*

	Causes of damage					
	Radiation		Accidents on the production line;	Non-radiational etiological diseases		
Type of damage	Personnel in mines	Population due to station waste	extraction and transportation of coal	Personnel during extraction & treatment of coal	Population due to station waste	All causes
No. of cases of premature death	0.09	0.044	3.42	5.51	290	299
No. of cases of disablement	0.09	0.044	130.2	10.76	290	431
Reduction of life expectancy (years)	0.9	0.44	492.5	55.1	8700	9249
Losses in work capacity (years)	0.45	0.22	2021	107.6	5800	7929

TABLE 5.10 *Cancer-induced mortality attributable to radiation from all stages of the nuclear fuel cycle (deaths per GW(e) produced)*

Stage in the fuel cycle	Personnel	Population	Total
Extraction of uranium ore	0.015	0.0	0.015
Treatment of ore and manufacture of fuel	0.01	0.0	0.01
Receipt of energy at plant	0.1	0.015	0.115
Regeneration of fuel	0.12	0.044	0.164
Dismantling of reactor	0.003	0.0	0.003
Research investigations	0.05	0.001	0.051
All stages	0.3	0.06	0.36

comparison between sources of power and causes of death. The amounts of electricity generated from the four chief sources of energy are shown in Table 5.11 (Hamilton and Manne, 1977).

3 Comparative Assessments and the Exposure Commitment Method

It was stated at the outset that the dose commitment concept was developed to assess total effects of released radioactive material in the environment. This has been developed as the exposure commitment method for assessing the effects of other environmental pollutants. This methodology allows comparative environmental impacts to be displayed. As an example of this, the results presented by UNSCEAR that compare the collective dose commitments from

TABLE 5.11 *Estimated health effects in 1975 associated with production of electric power in the U.S.A.*

Fuel	KWh(e) × 10[9a]	Equivalent no. 100-MW(e) plants	Estimated deaths	Estimated disabilities
Coal	844	128	1 900–15 000	25 000–39 000
Oil	292	44	88– 4 000	4 000– 7 000
Gas	297	45	6	600
Nuclear	168	26	18–42	130–470
TOTALS	1 604	243	2 000–19 000	29 000–48 000

[a] Preliminary

nuclear power with those of other sources of radiation (Butler and Hyslop, 1980) are of interest.

It is necessary to develop and particularly to apply the method on the basis of a unit measure of a given practice in order to provide a rational approach to the comparative assessment of environmental impacts of coal combustion and conversion, and other fuel cycles.

4 References

BARRY, P.J. (1979). *An Introduction to the Exposure Commitment Concept with Reference to Environmental Mercury.* Report no. 12. Monitoring and Assessment Research Centre, London.

BENNETT, B.G. (1981). The exposure commitment method in environmental pollutant assessment. *Environmental Monitoring and Assessment*, **1**, 21–36.

BREWER, S.T. (1976). Quantification and comparison of external costs of nuclear and fossil electrical power systems. *Energy and the Environment–Cost–Benefit Analysis* (ed. by R.A. Karam and K.Z. Morgan). Energy, Suppl. 1, 506–30. Pergamon Press, New York.

BUCHANAN, B.G. and BENNETT, B. G. (1980). *Relevance of the Commitment Method in Environmental Monitoring and Pollutant Assessment.* Report of the Monitoring and Assessment Research Centre, London.

BUTLER, G.C. (1978). Estimation of doses and integrated doses. *Principles of Ecotoxicology*, SCOPE Report 12 (ed. by G.C. Butler), pp. 91–112. John Wiley, Chichester.

BUTLER, G.C. and HYSLOP, C. (1980). Radioactive contamination of the environment. *Handbook of Environmental Chemistry* (ed. by O. Hutzinger). Springer-Verlag, Heidelberg.

BUTLER, G.C., KAROL, I.L., LINDELL, B.G., STEVENS, D.J. and ZELENY, V. (1972). Assessment and control of environmental contamination: experience with artificial radioactivity. *Biological Conservation*, **4**, 177–84.

CHATTERJEE, B., HOTZL, H., ROSNER, G. and WINKLER, R. (1980). Untersuchungen über die Emissionen von Radionukliden aus Kohlenkraftwerken. GSF – Bericht, S–617.

HALBRITTER, G., BRAÜTIGAM, F.W., FLUCK, F.W. and LESSMANN, E. (1981). Contribution to a comparative environmental impact assessment of the use of coal and nuclear energy for electricity generation for selected German site conditions. *International Symposium on Health Impacts of Different Sources of Energy.* Nashville, Tennessee.

HAMILTON, L.D. (1980). Comparative risk assessment in the energy industry. Proc. Symp. on *Health Risk Analysis.* Gatlinburg, Tennessee.

HAMILTON, L.D. and MANNE, A.S. (1977). Health and economic costs of alternative energy sources. *International Conference on Nuclear Power and Its Fuel Cycle.* IAEA-CN-36/448. IAEA, Vienna.

ICRP (1976). *Recommendations of the International Commission on Radiological Protection.* ICRP Publication 26. Pergamon, Oxford.

ICRP (1979). *Limits for Intakes of Radionuclides by Workers.* ICRP Publication 30 (supplement to Part 1), Annals of ICRP **3**.

INTL. SYMP. (1978). International Symposium on General Air Pollution and Human Health with Special Reference to Long-term Effects. *Environmental Health Perspectives*, **22**, 1–123.

KELLY, G.N. (1975). *The Predicted Radiation Exposure of the Population of the European Community Resulting from Discharges of Krypton–85, Tritium, Carbon–14 and Iodine–129 from the Nuclear Power Industry to the Year 2000.* Commission of the European Communities, Luxembourg.

KOLB, W. (1979). Radioaktive Stoffe in Flugaschen aus Kohlenkraftwerken. PTB – Mitteilungen **89**, 77–82.

LINDELL, B. (1960). An approach to the question of computing doses and effects from fall-out. *Health Physics*, **2**, 341–65.

LINDELL, B. (1978). Source-related detriment and the commitment concept: applying the principles of radiation protection to non-radioactive pollutants. *Ambio*, **7**, 250–9.

MCBRIDE, J.P., MOORE, R.E., WITHERSPOON, J.P. and BLANCO, R.E. (1977). *Radiological Impact of Airborne Effluents of Coal-fired and Nuclear Power Plants.* ORNL–5315, Oak Ridge National Laboratory, Tennessee.

MINISTRY FOR RESEARCH AND TECHNOLOGY (1979). *German Risk Study for Nuclear Power Plants.* TUV Rheinland, Cologne.

NATUSCH, D.F.S. (1978). Potentially carcinogenic species emitted to the atmosphere by fossil-fueled power plants. *Environmental Health Perspectives*, **22**, 79–90.

NISHIWAKI, Y., TSUNETOSHI, Y., SHIMIZU, T., UEDA, M., NAKAYAMA, N., TAKAHASHI, H., ICHINOSAWA, A., KAJIHARA, S., OHSHINO, A., OGINO, M. and SAKAKI, K. (1971). Atmospheric contamination of industrial areas including fossil-fuel power stations, and a method of evaluating possible effects on inhabitants. *Environmental Aspects of Nuclear Power Stations*, pp. 247–78. STI/PUB/261. International Atomic Energy Agency, Vienna.

O'BRIEN, B.J. (1979). *The Exposure Commitment Method with Application to Exposure of Man to Lead Pollution.* Report no. 13. Monitoring and Assessment Research Centre, London.

ULMER, W.T. (1977). Wirkung von SO_2 in der Atemluft. *VGB-Kraft-werkstechnik*, **57**, Heft 8, August.

UNEP (1981). *Environmental Impacts of Production and Use of Energy*, Vol. 1. Study Director E. El-Hinnawi. Tycooly International, Dublin.

UNSCEAR (1977). *Sources and Effects of Ionizing Radiation.* Official Records of the General Assembly, Thirty-second Session Suppl. No. 40 (A/32/40). United Nations, New York.

USDOE (1979). *Technology for Commercial Radioactive Waste Management.* DOE/ET – 028, U.S. Department of Energy, Washington.

WILSON, R. (1978). Risks caused by low levels of pollution. *The Yale Journal of Biology and Medicine*, **51**, 37–51.

WILSON, R. and JONES, W.J. (1974). *Energy, Ecology and the Environment.* Academic Press, New York.

6

The Environmental Significance of Trace Elements from Coal Combustion and Conversion Processes

1 Occurrence and Distribution of Trace Elements in Coal

1.1 The origin of trace elements in coal

The large numbers of elements present in coal are derived from several processes which took place before, during and after its formation. Coalification itself consists of two distinct stages (Firth, 1971). The first involves formation of peat from the carbonaceous flora by bacterial and fungal action, resulting in the production of large amounts of humic acids and chelating agents, which can efficiently take up cations. The covering of the peat layer by sediments marks the beginning of the second stage, in which biological processes are replaced by the action of pressure, temperature and time.

Some of the original trace elements, taken up by the plant species of the carbonaceous swamps, will have been retained after their breakdown and incorporated into the developing coal matter. Contemporaneous sedimentation may also have introduced high concentrations of minerals, with humic acids from the decaying flora solubilizing and binding these elements. Trace elements will also have been gained from and lost to inflowing ground-water during and after the coalification process. Finally, migration of the metal-rich magmatic fluid to areas adjacent to coal deposits can result in epigenic mineralization, as has occurred in the South Wales Coalfield (Firth, 1971).

1.1.1 The chemical association of trace elements with coal

Trace elements in coal may be classified according to whether they are primarily associated with the organic fraction of coal, with the mineral fraction or with both. Ruch, Gluskoter and Shimp (1974) examined the "organic affinity" of trace elements in various coals by analysis of specific gravity fractions. Most of the elements of environmental concern, including As, Cd, Hg, Pb and Zn were concentrated in the inorganic fraction of coal. Some, such as Cr, Cu and Se were present in both the mineral and organic matter, but those also tended to be inorganically associated; B and Be were found to be largely combined with the organic fraction of coal. In view of these findings, Ruch, Gluskoter and Shimp (1974) suggested that those elements with organic affinities were probably inherited from an early period of coal swamp formation. On the other hand, those in inorganic association, originated from emplacement by epigenetic mineralization, as they were often present in discrete minerals such as sulphides and carbonates.

1.2 Concentrations of trace elements in coal

1.2.1 Evaluation of reported values

Accurate values for the trace element content in coal are required before any assessment of their significance can be made. Prior to 1970, however, elemental measurements in coal employed high temperatures to remove the organic fraction before analysis (Clarke and Swaine, 1962; Zubovic, Sheffey and Stadnichenko, 1967; Abernathy, Peterson and Gibson, 1969). This method resulted in losses of volatile elements, an underestimate of their concentration and the production of unreliable data. More sophisticated analytical techniques have been developed in recent years and standard coal samples from the U.S. Bureau of Standards are available for inter-laboratory comparison (Ondov *et al.*, 1975).

Unrepresentative trace element levels in coal may also be found in those geological studies concerned with the exploitation of coal as a commercial source of certain elements. In such reports, only unusually high values were selected for publication (see for example, Bethell, 1962).

In an attempt to obtain a true picture of the variation in trace element content of coal for a particular region, studies such as the Illinois Geological Survey (Ruch, Gluskoter and Shimp, 1974) have analysed large numbers of samples collected from the same seam. The results of this investigation, summarized in Table 6.1, revealed that the elements As, Cd, Hg, Pb and Zn each displayed a highly skewed distribution, with very large ranges and standard deviations. A second group, comprising B, Cr, F, Ni, Se and V were approximately normally distributed with smaller ranges and standard deviations.

1.2.2 Variation in trace element content

Reported concentrations of trace elements from different coals often display considerable variability. It would be desirable to

TABLE 6.1 *Concentrations (μg g^{-1} dry weight) of trace elements for 82 coals from the Illinois Basin (Ruch, Gluskoter and Shimp, 1974)*

Element	Mean	Standard deviation	Range
Antimony (Sb)	1.35	1.42	0.2 – 8.9
Arsenic (As)	14.9	18.9	1.7 – 93.0
Beryllium (Be)	1.72	0.83	0.5 – 4.0
Boron (B)	113.8	51.7	12.0 – 224.0
Cadmium (Cd)	2.89	8.32	0.1 – 65.0
Chromium (Cr)	14.1	7.48	4.0 – 54.0
Fluorine (F)	59.3	19.8	30.0 – 143.0
Lead (Pb)	39.8	45.9	4.0 – 218.0
Mercury (Hg)	0.21	0.22	0.03 – 1.6
Nickel (Ni)	22.4	10.8	8.0 – 68.0
Selenium (Se)	1.99	0.93	0.45 – 7.7

ascertain whether this variability could be explained by or related to the rank of coal or its geographical location.

Coal rank

In view of the literature available at the time, Bethell (1962) concluded that trace element content generally decreased with increasing rank of coal. Although there have apparently been no recent assessments concerned specifically with this issue, there is little evidence available to support the above suggestion. Indeed, in the South Wales coalfield, it is the high-rank, anthracite coals which contain the highest concentrations of trace elements (Chatterjee and Pooley, 1977).

Table 6.2 presents trace element data, taken from Schwitzgebel *et al.* (1975) for three coals from the U.S.A.; a lignite from North Dakota and two sub-bituminous samples, both from Wyoming. With the exception of As and B, concentrations of trace elements in the low-grade lignite are similar or actually lower than in the two sub-bituminous coals from the same state. Table 6.2 also gives the elemental concentrations for another lignite, from the F.R.G. (Heinrichs, 1977). Although only a limited number of elements were determined, it is apparent that the values obtained are much lower than in either of the two coal ranks in the American study.

It should be noted that low-grade coal deposits have been associated with elevated concentrations of both As and U in two

TABLE 6.2 *Trace element content (μg g⁻¹ dry weight) in coals of different rank from the U.S.A. and F.R.G.*

Element	Coal rank: Lignite (N. Dakota)[a]	Sub-bituminous (Wyoming)[a] 1	Sub-bituminous (Wyoming)[a] 2	Lignite (F.R.G.)[b]
As	7.8	0.8	2.7	ND
B	146.0	49.0	32.5	ND
Cd	0.2	0.2	0.1	0.02
Cr	12.7	21.0	9.2	3.2
F	54.0	138.0	68.0	ND
Pb	0.8	4.2	2.4	1.1
Hg	0.07	0.13	0.14	0.05
Ni	5.5	8.5	2.2	1.3
Se	1.3	2.2	1.7	ND
U	1.5	1.0	1.3	ND
V	15.0	49.0	20.8	4.2
Zn	7.5	24.0	4.2	2.0

ND = Not determined.
[a] Schwitzgebel *et al.* (1975).
[b] Heinrichs (1977).

specific areas of the world. It has been known for some time that certain lignite deposits in the western USA contain high concentrations of U and Th (Denson, 1959). However, the average figure of 25 $\mu g\ g^{-1}$ U for all western lignites, given by Smith (1977), requires substantiation. In particular areas, these lignite deposits do certainly contain high concentrations of U; in Willeston Basin, Barber and Giorgio (1977) report a value of 0.18%. In spite of the environmental implications, it is not known whether these uraniferous coals are being utilized for combustion on any large scale.

In the case of As, particular lignite deposits in Czechoslovakia contain from 900 to 1500 $\mu g\ g^{-1}$; serious As contamination has arisen in the vicinity of the power plant which consumes this coal (Bencko and Symon, 1977).

Geographical location

In the U.S.A., where the most comprehensive studies have taken place, the trace element content of coal is known to vary considerably

TABLE 6.3 *Average trace element concentration* ($\mu g\ g^{-1}$) *in coal by state in the U.S.A. (Dvorak et al., 1977)*

Element	N. Appalachian		S. Appalachian				Eastern interior			Powder River region		Four Corners	
	W.Va.	Penn.	Eastern Ky.	Ala.	Va.	Tenn.	Ill.	Western Ky.	Ind.	Mont.	Wyo.	New Mexico	Ariz.
As	9	16	6	13	10	9	6	7	7	6	1	2	2
B	20	15	19	30	13	24	81	70	85	60	36	43	49
Cd	–	–	–	–	–	–	2.9	–	–	0.1	0.46	<2.6	<0.54
Cr	19	24	19	19	20	19	29	18	19	3	5.8	11	9.7
Cu	11	13	11	14	13	11	8.3	8.8	9.7	3.2	4.4	5.9	4.9
F	70	90	30	90	50	120	59	–	50	70	160	160	70
Pb	4.9	5.2	4	3.7	6.1	4.9	33	6.4	7.2	4.8	0.61	4.7	3.9
Hg	0.12	0.20	–	–	–	–	0.18	–	0.08	0.07	0.05	0.08	0.05
Ni	18	20	16	17	22	16	25	16	33	3.3	4.1	8.1	4.9
Se	3.4	3.7	3.1	5.1	4.4	4.9	2	3.1	4	3	0.8	2	2.1
V	30	33	29	31	33	34	35	32	35	12	15	25	9.7
Zn	17	22	15	22	23	23	140	48	73	42	37	19	9.7

TABLE 6.4 *Trace element content ($\mu g\ g^{-1}$) of British and Australian coals*

	Australian bituminous coals[a]		British coals[b]	
Element	Mean	Range	Mean	Range
As	3	>1 – 55	18	4 – 40
B	60	1.5 – 300	NG	NG
Cd	0.10	0.05 – 0.2	0.4	0.3 – 0.8
Cr	6	1.5 – 30	33.6	12 – 50
F	80	15 – 500	114	98 – 130
Hg	0.10	0.03 – 0.4	0.5	0.4 – 0.6
Ni	15	0.8 – 70	27.9	12 – 40
Pb	10	1.5 – 60	38	28 – 60
Se	0.79	0.21 – 2.5	2.8	1.8 – 4.4
U	2	0.4 – 5	1.3	0.5 – 2.3
V	20	4 – 90	76	38 – 134
Zn	25	12 – 73	NG	–

NG = Not given.
[a] Swaine (1977, 1979).
[b] Gibson (1979).
No sample size available for either study.

between regions. Table 6.3 assigns values to the main coal-bearing states; the large range encountered makes it difficult to assess regional patterns or the national average; As, however, would appear to decrease in concentration from eastern to western coalfields. In passing, it is encouraging to note that the analytical values for Illinois in Table 6.3 are similar to those found by Ruch, Gluskoter and Shimp (1974).

National averages have been published for both Australian and British coals (Table 6.4). Compared with coals from the U.S.A. and the U.K., Australian coals are generally lower in trace elements. Swaine (1977) suggests this is due to the low levels of pyrite and other sulphide minerals in these coals; certainly elements such as As, Cu, Ni and Pb are known to be found as sulphides in nature including some coals (Swaine, 1977).

1.2.3 Estimation of average trace element content and enrichment with respect to crustal abundance

Table 6.5 presents an estimated average trace element content of U.S. and Australian coal. The figures given have been assessed after

TABLE 6.5 *Estimated average trace element content of shale, coal and coal enrichment factor relative to crustal abundance*

Element	Mean value shale[a] (μg g^{-1})	Mean value coal[b] (μg g^{-1})		Enrichment factor[c]	
		U.S.A.	Australia	U.S.A.	Australia
As	13	15	3	8.3	1.7
B	100	50	60	5	6
Cd	0.3	1.3	0.1	6.5	0.5
Cr	90	15	6	0.2	0.1
Cu	45	19	1.5	0.3	0.03
F	740	74	80	0.1	0.1
Hg	0.4	0.18	0.1	2.3	1.3
Ni	68	15	15	0.2	0.2
Pb	20	16	10	1.3	0.8
Se	0.6	4.1	0.79	82	16
Th	12	4.7	2.7	0.7	0.4
U	3.7	1.8	2	1	1.1
V	130	20	20	0.1	0.1
Zn	95	39	25	0.6	0.4

[a] Turekin and Wedepohl (1961).
[b] Swanson *et al.* (1976); Swaine (1977).
[c] Calculated as the ratio of the concentration in coal compared to its average crustal value (Mason, 1966).

considering the results of studies already discussed and also those of Bertine and Goldberg (1971), Gladney (1974), Swanson *et al.* (1976), Coles, Ragaini and Ondov (1978) and Van Hook (1978). These values have also been compared to the average concentration of each element in the earth's crust, using the figures of Mason (1966). This ratio, known as an enrichment factor, is a simple way of assessing whether the element is enriched or depleted in coal compared to its crustal abundance. It is apparent from Table 6.5 that the trace elements given may be divided into three groups, according to whether they show enrichment, depletion or are actually present at similar concentrations to the earth's crust. Five elements show enrichment factors greater than one; these are As, B, Cd (U.S. coals), Hg and Se; Pb and U show no significant trend of enrichment or depletion, while all others are depleted in coal compared to the crustal value.

2 Mobilization of Trace Elements by Coal Utilization

2.1 Conventional Coal-fired Power Plants

Although potentially toxic elements are generally present at relatively low concentrations in coal, significant mobilization still occurs, because of the very large quantities of coal consumed. The annual world production of coal in 1978 was estimated to be 2.974×10^9 tonnes (Quenon, 1980). A single large power plant (>2000 MW) will burn as much as 10^7 tonnes of coal a year (Wiersma and Crockett, 1978). In addition, the combustion process may produce physicochemical alterations to trace elements and their compounds which enhance their toxicity.

Coal combustion in modern electric power plants takes place in furnaces operating at about 1600°C. When coal is burnt, a portion of the non-combustible mineral matter will be retained in the furnace as either slag or bottom ash, while the rest will leave as fly ash with the flue gases and volatilized minerals. A further fractionation occurs when the flue stream passes the particulate control devices. These may remove up to 99% of the larger fly ash particulates but are less efficient in collecting the smaller (<1 μm) particles and vapour. The proportion of bottom ash, slag and fly ash produced will vary not only with the ash content of coal, but with the plant design and the particulate collection device employed. For example, ash removed in a wet condition may have considerably higher elemental enrichment. In cyclone-fed units the total ash produced is divided equally between bottom ash and fly ash, while in the more commonly occurring pulverized coal plants, over 90% of the total ash may be fly ash (Bolton *et al.*, 1973).

2.1.1 Behaviour of trace elements within power plants

Several comprehensive investigations into the behaviour of trace elements in power plants have involved a detailed examination of the partitioning of trace elements between the various exit streams, to allow an assessment of their mass-flow rates. Such mass-balance studies have been carried out for a whole range of trace elements (Gladney, 1974; Kaakinen *et al.*, 1975; Klein *et al.*, 1975; Schwitzgebel *et al.*, 1975) and also for single elements (Billings and Matson, 1972; Andren, Klein and Talmi, 1975). The results obtained have been explained in relation to the chemical properties of trace elements and their compounds and to the operating conditions within the power plant.

Partitioning of elements

Mass-balance studies reveal that coal combustion in power plants produced a consistent and selective partitioning of trace elements between the various exit streams, permitting the classification of elements into three groups:

Group I - These elements are partitioned equally between bottom ash and fly ash and are not concentrated in the outlet fly ash. Elements in this group include Al, Ba, Ca, Ce, Cs, Fe, K, Mg, Mn and Th.
Group II - These elements are concentrated in the inlet fly ash compared to the bottom ash and the outlet fly ash compared to the inlet fly ash. This group consists of As, Cd, Cu, Pb, Sb, Se and Zn.
Group III - Elements in this group are mainly present in the gas phase and are depleted in all ashes. This group contains three elements, Br, Hg, and I.

Several elements, Cr, Ni, U and V, exhibit intermediate behaviour between Group I and Group II.

This classification may also be extended to the concentration of trace elements as a function of the particle size of fly ash. Natusch, Wallace and Evans (1974) reported that concentrations of elements in Group II increase as the particle size decreases, while Group I elements showed no such relationship, being present in roughly equal concentrations in the various size fractions of fly ash.

An enrichment mechanism

Natusch, Wallace and Evans (1974) explained the enrichment of certain trace elements in smaller particles by a volatilization-condensation mechanism; this also accounts for the observed difference in partitioning between the various exit streams. The model suggests that when coal is burnt, elements from Group II are volatilized and will thus be depleted from the slag and bottom ash. As the flue gases cool, these elements will preferentially condense out on the smaller fly ash particles because such particles have a larger surface area to volume ratio. Elements in Group I will remain condensed at the temperature of coal combustion and form the bottom ash and the fly ash particles on which the volatile elements condense. The selective retention of coarser fly ash particles by the collection devices will further enhance the small particle association of volatile elements. Group III elements, being very volatile, will mostly remain in the gas phase during passage through the power plant, although they may be, in part, associated with fly ash.

Although the manner in which trace elements are partitioned is related to their boiling point, their chemical form in coal also plays an important role. Both Kaakinen *et al.* (1975) and Klein *et al.* (1975) point out that all those elements in coal classified in Group I are lithophiles in crustal rocks, being associated with aluminosilicate minerals. Conversely, elements in Group II are chalcophiles, as they are present as sulphide minerals in crustal components. Klein *et al.* (1975) suggest that when coal is burnt, the lithophile elements remain condensed, but chalcophiles are decomposed in the reducing conditions of the combustion zone, forming volatile species which are vaporized.

The combustion characteristics of U illustrates the need to consider the chemical composition of trace elements in coal, in order to understand their behaviour. Although U displays little enrichment on fly ash compared to bottom ash, it is associated with the smaller fly ash particles (Coles, Ragaini and Ondov, 1978) and is thus intermediate between Group I and Group II elements. Uranium is known to exist in coal as both the silicate mineral, coffinite, and uraninite (UO_2). Upon combustion, the refractory coffinite will remain in the slag and bottom ash, but the uraninite will be vaporized, condensing on fly ash particles when the flue gases cool. Furthermore, the silicate content of coal may be an additional factor in the combustion chemistry of U, as this coal determines the relative proportions of the two forms of U and this affects its combustion behaviour. The partitioning of other trace elements may also be affected by the content of major elements in coal. Kaakinen *et al.* (1975) for example, report that As appears to be retained in ashes of high Ca coals, while in low Ca coals it is more efficiently volatilized.

Enrichment of trace elements in fly ash

The extent of trace element enrichment on fly ash is most conveniently expressed as an enrichment factor (EF). This is the ratio of the concentration of an element (X) to that of Al, or another refractory element, in the fly ash, divided by the corresponding abundance ratios in crustal rocks (Mason, 1966).

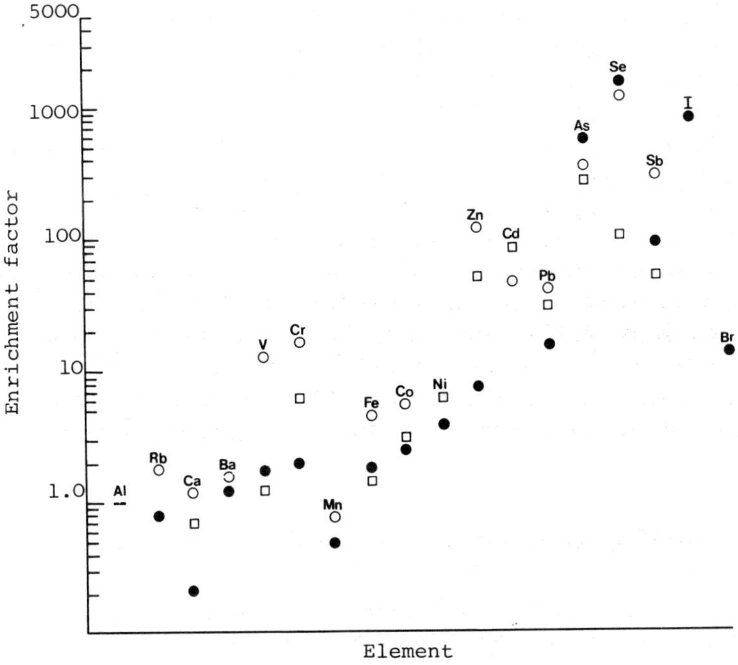

FIGURE 6.1 *Elemental enrichment factors of suspended fly ash particles relative to crustal abundance from three U.S. power plants:* ●, *Chalk Point (Gladney, 1974);* ○, *Allen Steam Plant (Klein et al., 1975);* □, *unnamed facility (Natusch, Wallace and Evans, 1974).*

$$EF = \frac{([X] / [Al])\ \text{fly ash}}{([X] / [Al])\ \text{crust}} \quad .$$

This normalization technique takes into account the apparent enrichment of fly ash caused by carbon losses in the combustion of coal.

Figure 6.1 shows trace element enrichment factors relative to crustal abundance, of suspended fly ash particles collected from three coal-fired power plants in the USA: Chalk Point (Gladney, 1974; Gladney *et al.*, 1976); the Allen Steam Plant (Klein *et al.*, 1975) and an unnamed facility (Natusch, Wallace and Evans, 1974). The values found are generally comparable at the three installations, with significant enrichment factors for As, Cd, I, Pb, Sb, Se and Zn. The Allen Steam Plant did, however, produce greater enrichments than

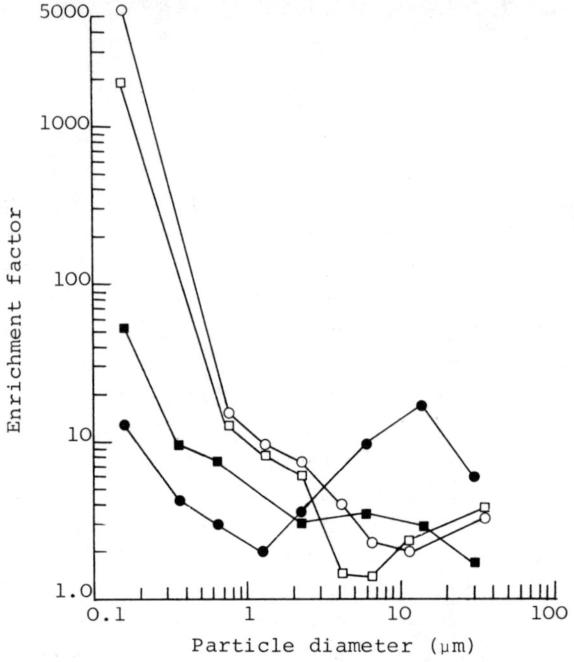

FIGURE 6.2 *Enrichment factors of As and Se with respect to coal from two U.S. power plants, expressed as a function of particle size: ●, Se; ■, As (Gladney et al., 1976). ○, Se; □, As (Ondov, Ragaini and Biermann, 1978).*

Chalk Point for Co, Cr, Fe, Pb, Sb and V. This difference may be explained by the higher combustion temperature employed at the Allen Plant, producing a greater vaporization of the above elements in the furnace. Such a suggestion is supported by the finding that As and Se showed greater EF values at Chalk Point; presumably the lower temperatures encountered at this plant allow a greater proportion of these two volatile elements to condense from the gas phase.

The extent of trace element enrichment on the finer fly ash particle is a matter of some importance, because it is these particles which can pass through the collection devices and are deposited efficiently in the pulmonary region of the lung. Natusch, Wallace and Evans (1974) were the first to report dramatic increases in the concentrations of several toxic elements, As, Cr, Ni, and Pb, on sub-micron fly ash particles. Unfortunately, these findings are difficult to evaluate in relation to later studies, as no information was provided on the elemental composition of the original coal. For this reason, no enrichment factors can be calculated for comparisons with other investigations.

In a more comprehensive study, Gladney *et al.* (1976) fractioned suspended fly ash particles with a cascade impactor and examined enrichment factors for a suite of elements. In contrast to Natusch, Wallace and Evans (1974), both Cr and Ni generally showed very little increased enrichment with decreasing particle size, although they do so in some Australian coals (Swaine, 1979). Furthermore, although enrichment factors for Pb and Sb increased on the finer particles, the extent of enrichment was much less than found by Natusch, Wallace and Evans (1974). The volatile elements, As and Se, also displayed an increased enrichment on smaller particles, but as Figure 6.2 shows there was also a definite minimum in the mid-sized (1–5 μm) range for Se. This bimodal behaviour cannot be explained by the simple volatilization-condensation model, discussed above, and recent work (Smith, Campbell and Nielson, 1979) indicates that little enrichment occurs below particle diameter of 1 μm.

The result of Gladney *et al.* (1976) must also be viewed with caution, however, as serious errors in using cascade impactors to estimate enrichment factors as a function of particle size in power plants were reported by Ondov, Ragaini and Biermann (1978). By use of scanning electron microscopy, the latter study found that those particles collected on the upper impactor stages were actually much smaller than indicated by the 50% cut-off diameters given for the impactor. Furthermore, on the back-up filters, large particles (>2 μm) constituted more than 90% of the total mass and 20–30% of volatile elements such as As and Se. Back-up filters are employed to collect the respirable, sub-micron particles, too small to be retained by impaction. In view of these findings, Ondov, Ragaini and Biermann (1978) suggested that enrichment factors on small particles may be more than an order of magnitude greater when the large particle component on the back-up filter is taken into account.

In Figure 6.2, enrichment factors of As and Se *vs.* particle size have been calculated from the data of Ondov, Ragaini and Biermann (1978), employing the elemental coal consumption given in a different study of the same power plant (Coles *et al.*, 1979). Values on the back-up filter have been corrected for bounce-off effects and are much larger than found by Gladney *et al.* (1976). Additionally, Se shows no evidence of the biomodal behaviour, reported by Gladney *et al.* (1976).

Campbell *et al.* (1978) carefully size-separated fly ash collected by

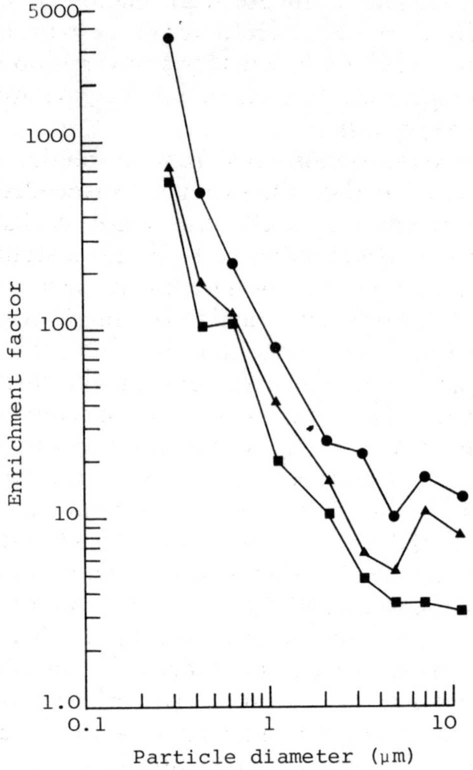

FIGURE 6.3 *Enrichment factors of selected elements with respect to coal, expressed as a function of particle size, from a domestic heating unit (Block and Dams, 1976):* ●, *Se;* ▲, *Zn;* ■, *As.*

an electrostatic precipitator, using a combination of an air centrifugation–elutriation technique; the accuracy of the fractionation was confirmed by scanning electron microscopy. Concentrations of As, Pb and Se showed a dramatic increase with decreasing particle size, with no evidence of bimodal behaviour. Additionally, and in contrast to Gladney *et al.* (1976), Cr, Ni and Zn also displayed small particle association. Unfortunately, it is again not possible to compare these results directly with other studies, as no information on the elemental content of coal was provided.

Clearly, further work is required in order to clarify the particle-size distribution of potentially toxic trace elements in suspended fly ash. Initially, scanning electron microscopy should be used to establish the reliability of the fractionation technique employed. Attention can then be directed to the extent of enrichment on sub-micron particles and to whether any variation in enrichment can be linked to specific differences in the operating conditions of coal-fired power plants.

If, as the volatilization-condensation model suggests, volatile elements are deposited on the surface of suspended fly ash particles, then such elements may well pose a greater hazard than simple chemical analysis would indicate. To date, no studies have examined the surface predominance of potentially toxic elements on sub-micron particles, presumably due to technical difficulties. Nevertheless, Linton *et al.* (1976) found that both Pb and Zn showed significant increases in concentration near the surface of 75–100 μm fly ash particles. The layer of surface enrichment appeared to be about 0.1 μm in depth. By extrapolation this would suggest that a large proportion of the trace element mass of a 1 μm particle would be present within this shell. However, although elements such as Na, C and S decreased with depth in both 1 μm and 2.6 μm particles (Campbell *et al.*, 1979), the relative thickness of the surface layers was similar in the two sized particles. To account for both the surface predominance of volatile elements and the smaller surface layer than predicted by the volatilization-condensation model, Campbell *et al.* (1979) proposed a mechanism involving the bursting of larger particles to form smaller fragments, with surface deposition of volatile elements occurring before, during and after the process.

2.2 Other Combustion Facilities

2.2.1 Domestic coal combustion

Coal burning in the domestic sector and away from centralized units has declined in recent years, but 15% of the total U.K. consumption was still used for this purpose in 1977–78 (Gibson, 1979).

Not surprisingly, the reported combustion behaviour of trace elements in a domestic unit corresponds to that in power plants, with enhanced concentrations of chalcophiles (Group II elements) on fly ash, which further increase with decreasing particles size (Block and Dams, 1976). What is surprising, however, was that in the absence of any particulate collection control device, over half the suspended fly ash particles from this domestic appliance were smaller than 1 μm in diameter. The significance of this mass distribution is revealed when enrichment factors as a function of particle size are calculated for several elements from the raw data of Block and Dams (1975, 1976). Figure 6.3 reveals that As, Se and Zn each show increasing enrichment as particle size decreases. Compared to power plants, the values found on sub-micron particles are much greater than reported by Gladney *et al.* (1976) but similar to those computed from the results of Ondov, Ragaini and Biermann (1978). One possible explanation for the high enrichment factors encountered in this domestic appliance is the low temperature (40°C) at the point of particulate sampling in the chimney; this may allow efficient condensation of volatile elements to occur *in situ*. In order to establish whether such enrichments on sub-micron particles are characteristic of domestic heating units in general, it is necessary to perform further investigations of other appliances.

2.3 *Coal Conversion Processes*

There are only a few commercial-sized gasification plants in existence and the only liquefaction plant presently operating on a commercial scale is the Sasol plant in South Africa. However, a number of governments are interested in commercial coal conversion processes and the U.S. Government was committed to the production of 2 million barrels of synthetic fuels per day by 1990 (Wade, 1979). It therefore seems desirable to attempt to assess the environmental acceptability of such processes and here the behaviour of trace elements in coal conversion plants is considered. As far as possible, information from commercial-sized conversion plants is evaluated (Bombaugh *et al.*, 1980; Patterson, 1980), but it has proved necessary to rely mainly upon results from bench-scale or pilot plant processes and extrapolation from conventional coal-fired power plants.

One factor which may be of general importance when evaluating the mobilization of trace elements from conversion processes is the choice of coal type to be converted. This is because a higher mineral content in coal, including high FeS_2 and As, has been reported to have a definite catalytic effect upon hydrogenation, the effect depending upon the precise quantity and composition of the mineral matter present (Mukherjee and Chowdhury, 1976). Thus, coals which may be selected for conversion because of this catalytic effect may contain elevated concentrations of those trace elements associated with the ash fraction (Ruch, Gluskoter and Shimp, 1974). It is of interest to note, for example, that the coals of North Assam display unusually high hydrogenation rates compared to other Indian coals and contain elevated concentrations of Cr (200 $\mu g\ g^{-1}$), Mo (200 $\mu g\ g^{-1}$), Ni (225 $\mu g\ g^{-1}$) and V (150 $\mu g\ g^{-1}$) (Mukherjee and Chowdhury, 1976).

2.3.1 Gasification

Gasification processes have been described in Chapter 4. The gas treatment steps following gasification-shift conversion, methanation and dehydration have features common to all processes.

The following assesses trace element behaviour for three gasification processes to illustrate the range of reaction conditions and effluent characteristics encountered.

The Synthane gasifier

An example of this process is the U.S. Bureau of Mines laboratory-scale gasifier (Oldham and Wetherold, 1977). In this two gas

TABLE 6.6 *Trace element content ($\mu g\ g^{-1}$) of feed coal and effluent streams from a laboratory synthane gasifier*

		Effluent streams			
Element	Feed coal	Filter fumes	Char	Tar	Condensate*
As	0.87	3.7	6.5	0.17	0.054
B	86	64	380	12	35.6
Cd	0.097	0.88	1.6	–	0.001
Co	14	17	95	0.09	–
Cr	170	47	240	7.1	<0.001
F	490	610	150	0.97	23.5
Hg	0.1	0.2	ND	1.2	0.014
Mo	15	21	14	0.31	ND
Ni	43	12	25	1.2	0.028
Pb	0.55	2.2	21	0.22	0.33
Se	5.3	6.4	17	0.02	0.78
Th	3	5.8	4.3	0.06	ND
U	1.4	5.6	5.4	0.01	0.031
V	100	44	190	0.21	0.29
Zn	25	11	100	0.48	3.9

ND = Not determined.
Data from Forney, 1974 (* Oldham and Wetherold, 1977).

TABLE 6.7 *Enrichment ratios (ER) in relation to crustal abundance for various elements of filter fines in a synthetic gasifier (Koppenaal et al., 1979)*

Non-enriched elements		Enriched elements	
Element	ER	Element	ER
Na	2.0	K	3.1
Mg	2.7	Rb	6.5
Al	1.0	Co	4.1
Si	1.2	Ni	3.0
Ca	0.8	Zn	7.0
Ti	1.1	Cu	9.3
Zr	2.9	Ga	4.7
Fe	0.8	Ge	11.0
Cr	1.6	As	6.2
Be	2.4	Br	11.0
Sr	1.2	Sn	3.3
Ba	1.1	Sb	5.7
Hf	1.5	I	14.0
Gd	1.5	Mo	7.3
		Pb	3.8
		Cd	6.5

streams, one of them hydrogenated, are recombined, desulphurized and then methanated using a Ni catalyst.

The trace element content of several of the effluent streams from the laboratory-scale unit have been determined (Forney, 1974; Oldham and Wetherold, 1977) and are summarized in Table 6.6. Gasification char is the major sink for most of the elements in the feed coal and this has been confirmed recently (Koppenhaal *et al.*, 1979) in a study in which sub-bituminous feed coal was used. This work has also indicated, however, that high enrichment of trace metals takes place on filter fines through a volatilization-condensation mechanism similar to that observed in coal combustion (Table 6.7). Furthermore, as the gasifier char is intended to be burnt in an auxiliary utility boiler to provide steam and electricity for the gasification unit, particulate control devices will have to be employed in the utility boiler and an investigation into trace metal emissions resulting from char combustion carried out.

Although the condensate, formed after cooling the raw synthesis gas, contains low concentrations of most trace elements some enrichment of As, B, Hg, Sb, Pb and Se on fine particles is observed (Craun and Massey, 1979). Process waters, however, also contain significant amounts of As, Cd, Hg, Pb, Se and Sb and particularly high concentrations of Se were found in the condensate process water and tars. Mercury was not determined but its behaviour would be expected to resemble Se. In addition, relatively high levels of B and F were present in a condensate sample examined by Forney (1974).

The Koppers-Totzek gasifier

Information on this process is based on a semi-commercial plant with capacity for 3×10^7 g coal hr^{-1} (Oldham and Wetherold, 1977) and the theoretical calculations by Anderson, Hill and Fleming (1979) which computes volatility-condensation behaviour of volatile elements from thermodynamic properties.

The Koppers-Totzek process is regarded as relatively "clean" compared with other conversion techniques (Oldham and Wetherold, 1977). This is certainly borne out by the comparative data that have been calculated (Anderson, Hill and Fleming, 1979) and presented in Table 6.8. The higher temperature employed in the Koppers-Totzek gasifier will presumably result in a more complete vaporization of moderately volatile trace elements from the coal. For example, Pb shows a much higher volatility in the Koppers-Totzek process than in the Lurgi. The form and distribution of certain elements in raw and quenched gasifier product gas is shown in Table 6.9.

The Lurgi gasifier

No information regarding the trace element content of effluent streams from Lurgi gasifiers operating on a commercial scale was available until data on the Kosovo plant in Yugoslavia were published recently (Bombaugh *et al.*, 1980). Calculations indicated that the important streams would be the gasifier ash, the process condensate and the ash quench water (Table 6.9). In view of the operating conditions employed, it was suggested (Anderson, Hill and Fleming, 1979) that there would be a less efficient vaporization of trace elements compared to the Koppers-Totzek process, resulting in greater retention of these elements in the gasifier slag or ash (Table 6.9). The recent study (Bombaugh *et al.*, 1980) of the Lurgi Kosovo Kombinant plant in Yugoslavia provides a considerable amount of data on stream characterization of a commercial plant in normal operation; this is presented in Chapter 8 (Table 8.7).

TABLE 6.8 *Calculated partitioning of Hg, Se, As, B, and Pb for various streams in the Koppers-Totzek and Lurgi processes (Anderson, Hill and Fleming, 1979)*

Element	Hg		Se		As		B		Pb	
Concentration in coal (ppm)	0.2		2.08		14		10.2		35	
Process	Koppers-Totzek	Lurgi	Koppers-Totzek	Lurgi	Koppers-Totzek	Lurgi	Koppers-Totzek	Lurgi	Koppers-Totzek	Lurgi
Input streams (kg day^{-1})										
Coal feed	3.672	3.672	38.3	38.3	257.96	257.96	188.3	188.3	641.2	641.2
Output streams (kg day^{-1})										
Discharge ash	neg.	neg.	neg.	19.1	neg.	128.98	neg.	94.1	neg.	641.2
Solids from waste-water treatment	neg.	neg.	neg.	neg.	257.96	100.06	188.3	94.1	641.2	0.005
Product sulphur	0.0007	3.635	38.0	19.1	neg.	28.73	neg.	neg.	neg.	neg.
Sulphur recovery tail gas	neg.	neg.	neg.	neg.	neg.	neg.	neg.	neg.	neg.	neg.
CO_2 vent gas (I)	0.0015	0.0092	<0.4	<0.2	neg.	<0.275	neg.	neg.	neg.	neg.
Methanation quench	neg.	neg.	neg.	neg.	neg.	neg.	neg.	neg.	neg.	neg.
CO_2 vent gas (II)	0.0015	0.0173	neg.	neg.	neg.	neg.	neg.	neg.	neg.	neg.
Glycol cooler recovery	3.394	neg.	neg.	neg.	neg.	neg.	neg.	neg.	neg.	neg.
Product SNG	0.27	0.0097	neg.	neg.	neg.	neg.	neg.	neg.	neg.	neg.

neg. = negligible.

TABLE 6.9 *Form and distribution of selected trace elements (g-mol g-mol^{-1}) in raw and quenched gasifier product gas in the Koppers-Totzek and Lurgi processes (Anderson, Hill and Fleming, 1979)*

	Lurgi (20 atm operating pressure)	Koppers-Totzek (2 atm operating pressure)
Raw gasifier product gas		
AsH_3	1.71×10^{-7}	2.47×10^{-15}
As_4	1.56×10^{-7}	9.56×10^{-12}
As_2	3.20×10^{-9}	1.20×10^{-6}
H_2Se	1.36×10^{-7}	3.68×10^{-7}
$B(OH)_3$	4.88×10^{-5}	7.29×10^{-5}
PbS	8.13×10^{-14}	7.62×10^{-7}
$PbC1_2$	1.11×10^{-11}	1.30×10^{-12}
PbO	2.81×10^{-25}	2.31×10^{-9}
Pb	6.94×10^{-16}	1.67×10^{-6}
Hg	8.55×10^{-9}	1.28×10^{-8}
Quenched gasifier product gas		
AsH_3	3.19×10^{-7}	2.68×10^{-15}
H_2Se	2.50×10^{-7}	4.00×10^{-7}
Hg	1.58×10^{-8}	1.38×10^{-8}

Common auxiliary processes

Synthetic gas is subjected to shifting in all gasification processes, to produce the appropriate ratio of CO to H_2, prior to methanation. This is achieved in a catalytic reactor, employing a Cr-promoted Fe catalyst. Desulphurization units convert S to its elemental form. This process also employs catalysts, including those used in the Stretford-Beavon technique (Co-Mo and V) and possibly the Giamarco-Vetrocoke process (As). The methanation of synthetic gas is also carried out in all gasification processes, by use of a Ni catalyst. It is not known whether significant amounts of those potentially toxic elements employed in the above catalytic processes become entrained in the gas stream. Nickel carbonyl ($Ni(CO)_4$), a carcinogenic compound, forms in methanizer gas (Ghassemi, Crawford and Quinlivan, 1979). During continuous processing, temperatures well above those at which Ni carbonyl forms (>275°C) are employed, but during start-up and shut-down, when lower temperatures are experienced, some formation of Ni carbonyl may occur. All traces would need to be removed eventually from the SNG delivered to the consumer, as upon decomposition in burners problems would arise resulting in failure of the burners. A detailed analysis of low-Btu product gas from a Wellman-Galusha gasifier (Thomas, Trede and Page, 1979) revealed significant amounts of $Ni(CO)_4$ and iron carbonyl ($Fe(CO)_5$). This is obviously an area of potential environmental concern, as well as technical interest from the point of view of burner efficiency.

2.3.2 Liquefaction

A number of different coal liquefaction processes have been described in Chapter 4. Two are discussed here: the Synthoil liquefaction process and the Solvent Refined Coal process. Development of the Synthoil process has been discontinued. However, recently much data has been published on the stream composition in the Synthoil process (Yavorsky and Akhtar, 1974; Schultz *et al.*, 1977, 1978 and Lett *et al.*, 1977) together with data for the centrifuged liquid product. Much of this data is consistent with the information concerning the SRC II process. This, together with SRC I, H-coal and Exxon Donor Solvent (EDS) will be the most important processes in the U.S.A. For this reason data for the Synthoil process are given here.

The Synthoil liquefaction process

In the Synthoil process a slurry of finely ground coal and oil is supplied to a reactor that contains a fixed-bed catalyst. Turbulent conditions in the reactor ensure contact between the catalyst (cobalt molybdate), coal plus the coal-derived liquid (or carrier) and the hydrogen that is fed in. Hydrogenation and desulphurization reactions occur during relatively short residence times in the reactor (a few minutes only).

The concentration of six trace elements in all the streams of the Synthoil process are given in Table 6.10. There is also some evidence that As may be present at an elevated concentration in the final oil product (Oldham and Wetherold, 1977).

The extent of trace element contamination of the Synthoil product oil will not only be relevant to its commercial usage but will also be of importance in the liquefaction process itself, as the oil is intended to be used to provide heat for the plant. Presumably, particulate collection devices will be employed to reduce emissions from the

TABLE 6.10 *Trace element analysis of all streams of the Synthoil process in* $\mu g\ g^{-1}$ *(Schultz et al., 1977)*

Stream	Element					
	Cu	Cr	Mn	Ni	Cd	Pb
Feed coal	10	21	57	9.3	0.32	4.6
Recycle oil	4.7	7.7	12	7.4	0.083	1.4
Feed paste	6.6	13	29	9.1	.17	2.4
Scrubber influent	0.037	0.10	0.022	0.012	≤0.002	0.025
Gross liquid product	6.7	15	31	10	0.19	3.0
Centrifuged liquid product	2.7	7.6	11	6.6	0.077	1.1
Centrifuge residue	45	84	180	54	1.0	18
Scrubber effluent						
Aqueous	0.15	0.28	0.019	0.029	0.0006	0.18
Organic	0.12	0.17	≤0.004	≤0.004	≤0.014	≤0.5
Vapour knock-out traps						
Aqueous	20.1	0.20	0.25	0.45	0.03	1.2
Organic	15.9	0.20	0.30	0.90	0.04	1.2

combustion of this oil. If estimates of the trace element content are representative then the collected fly ash may contain elevated concentrations of several volatile elements.

In the Synthoil coal conversion process the catalyst is employed in the reactor vessel itself. It is thus possible that the oil products and the effluent streams will be contaminated with Co and Mo. Additionally, it is probable that replacement of this catalyst will have to be carried out frequently, as it will be in direct contact with large amounts of tars and oils in the reactor vessel. This may bring about disposal problems if the spent catalyst is not reclaimed.

Solvent-refined coal process - SRC I and SRC II

The primary product of the SRC process is a heavy solvent-refined coal, solid at ambient temperature, but a fuel gas and oil are also manufactured.

Table 6.11 shows the product yields for SRC I and SRC II processes and Table 6.13 gives trace element concentrations from SRC I process streams. The extent of the reduction in trace elements

TABLE 6.11 *SRC I and SRC II product yields as a percentage of moisture free coal*

Component	SRC I[a]	SRC II[b]	
H_2	−2.8	−3.1	
C_1–C_4 gases	5.9	11.0	
CO	0.02	0.2	
CO_2	1.4	1.9	
H_2O	5.0	9.8	
H_2S	1.7	2.8	
Light oil	2.4	6.9	Naphtha
Wash solvent	7.8	17.4	Middle distillate
Process solvent	−8.9	9.6	Heavy distillate
Solvent-refined coal	69.5	27.6	
Ash	11.9	11.5	
Insoluble organic material and coal	6.1	4.4	
TOTAL	100.0	100.0	

[a] Filby *et al.* (1978)
[b] Pittsburgh and Midway Coal Mining Co. (1978).

TABLE 6.12 *Trace element reduction in SRC compared with coal (Filby et al., 1978)*

Element	Equilibrium set 1		Equilibrium set 2	
	SRC/coal	Percentage reduction	SRC/coal	Percentage reduction
Ti	0.88	12	0.74	16
V	0.15	85	0.47	53
Ca	0.22	78	0.22	78
Mg	0.08	92		
Al(%)	0.02	98	0.03	97
Cl	0.61	39	0.34	66
Mn	0.60	40	0.40	60
As	0.16	84	0.07	93
Sb	0.08	92	0.04	96
Se	0.06	94	0.03	97
Hg	0.35	65	0.41	59
Br	1.70	+70	1.33	+33

TABLE 6.13 *Trace element concentrations in SRC I process streams in ppm except as noted (Filby et al., 1978)*

Element	Ground coal	SRC	Pyridine insolubles	Wet filter cake	Light oil	Process recycle solvent	Wash solvent	Sulphur	Process water	Effluent water
As	12.5	2.0	85.7	62.1	0.011	0.24	0.011	<2.0	0.006	<0.001
Sb	0.76	0.06	7.21	5.35	<0.4[a]	8.2[a]	<0.4[a]	<0.1	0.66[a]	2.0[a]
Se	2.0	0.12	16.5	11.3	51.6[a]	24.0[a]	14.4[a]	141.5	0.16	0.0012
Hg (ppb)	113.0	39.6	508.0	346.0	18.5	1.45	10.5	<100.0	106.0	3.2
Br	4.56	7.74	12.0	20.7	0.015	1.0	0.02	<3.0	15.6	31.8
Ni	14.9	<3.0	142.0	82.4	<0.03	0.4	<0.03	<28.0	<0.004	0.013
Co	5.88	0.22	40.7	26.5	<3.0[a]	40.7[a]	1.43[a]	110.0	0.2[a]	0.41[a]
Cr	13.7	1.64	106.0	69.2	37.3[a]	3590.0[a]	41.3[a]	<2.0	0.007	0.15
Fe (%)	2.11	0.03	16.8	11.7	2.90[a]	211.0[b]	11.2[a]	<0.1	0.30[a]	1.25[b]
Na	137.0	4.23	1020.0	623.0	0.60	0.50	0.45	3120.0	0.70	8.3
Rb	<4.0	<0.5	66.5	37.1	<0.01	0.02	<0.01	<9.0	0.78	0.52
Cs	0.75	0.02	5.08	3.2	1.06[a]	<1.2[a]	0.91[a]	<0.2	0.04[a]	0.02
K	1550.0	4.72	11100.0	6660.0	<0.1	0.25	<0.1	179.0	0.2	1.26
Sc	2.59	0.57	14.8	9.26	0.15[a]	32.8[a]	0.19[a]	<0.02	0.13	0.01[a]
Tb	0.39	0.045	2.06	1.34	0.13[a]	3.75[a]	<0.13[a]	<0.1	0.01[a]	0.01[a]
Eu	0.26	0.055	1.48	0.96	0.01	<0.01	<0.01	<0.01	0.01[a]	0.01[a]
Sm	2.62	0.29	16.9	8.16	<0.01	0.02	<0.01	0.61	0.08[a]	<0.06[a]
Ce	20.9	0.45	156.0	012.0	<0.004	<0.004	<0.003	<2.0	<0.2[a]	<0.2[a]
La	7.55	0.13	59.8	35.2	<0.01	0.01	<0.01	1.80	0.27[a]	0.5[a]
Sr	88.6	<6.0	456.0	453.0	<0.6	<0.2	0.74	<45.0	<0.01	<0.04
Ba	53.0	5.75	347.0	185.0	<0.1	1.14	<0.07	<39.0	<0.02	<0.04
Th	2.0	<0.22	12.8	7.70	<0.012	<0.001	<0.2	0.2	0.05[a]	<0.01[a]
Hf	0.51	0.084	3.30	2.20	<0.001	0.003	<0.001	<0.2	0.02[a]	<0.01[a]
Ta	0.14	0.046	0.71	0.42	<0.4[a]	2.53[a]	<0.3[a]	<0.2	0.02[a]	0.01[a]
Ga	3.56	1.79	19.4	11.3	<0.01	0.06	<0.01	<1.5	<1.0[a]	<4.0[a]
Zr	62.9	16.0	500.0	246.0	0.07	0.71	<0.1	<61.0	0.02	0.04
Cu	19.9	2.07	189.0	138.0	0.03	0.68	0.03	<1.0	<12.0[a]	<10.0[a]

[a] Values in ppb.
[b] Values in ppm.

TABLE 6.14 *Trace elements in SRC I and SRC II materials and in coal and fuel oil as ppm.*

Element	SRC I[a]	SRC II blend[b]	Illinois[a] coal	Residual fuel oil[c]
As	2.10	0.0148	16.3	0.055
Sb	0.136	0.00259	1.10	0.04
Se	0.123	0.0284	2.52	0.09
Hg	0.0432	0.00339	0.114	0.004
V	9.51	0.179	29.7	87.0
Ti	326.0	<10.0	595.0	<10.0
Ni	1.82	0.0992	13.7	12.5
Co	0.314	0.00304	5.54	0.32
Cr	4.19	2.57	14.3	0.07
Fe	438.0	4.14	22500.0	5.0

[a] Filby *et al.* (1978).
[b] Heavy/middle distillate ratio : 5.75.
[c] Cahili (1974).

in solvent-refined coal compared to the original coal used is shown in Table 6.12. It appears that the SRC residues (pyridine insolubles) may be a major trace element "sink" and the combination of high pressures and relatively low reaction temperatures may be responsible for the retention of volatile elements. Residues will be recycled to a gasification plant; if the Koppers-Totzek process is employed, the residue will be subjected to high temperatures in an unpressurized reactor vessel. The possibility thus exists that those volatile elements enriched in the residue will be efficiently vaporized under the conditions employed in the gasification unit.

Table 6.14 compares trace elements in SRC I and SRC II materials. It will be seen that particularly in the SRC II process where increased residence time and H_2 pressure increase conversion to liquids, the elements Hg, Se, As and Sb exhibit volatile behaviour. Transport in gaseous streams has been demonstrated (Filby, Khalil and Hunt, 1980). High concentrations of these elements, and other volatile metals, are also observed in process waters (Table 6.13) and some light oils (Table 6.15). Treatment processes that reduce toxic metals in effluent waters produce biotreatment sludges that contain high concentrations of Hg, Se, As and other trace elements. These sludges will present significant disposal problems as will the deposits that foul process units. This could result in costly shut-downs for removal

TABLE 6.15 *Trace elements in SRC I and SRC II distillate fractions as ppb except as noted*

Element	Separator No. 2 oil[a]	Separator No. 3 oil[a]	Solvent acc.oil[a]	Light oil SCR I[b]
As	57.7	32.9	12.7	2.85
Sb	4.45	<0.2	0.55	1.86
Se	214.0	57.1	115.0	23.0
Hg	776.0	4.6	27.3	3.4
Br	11.1	8.71	6.65	18.0
Ni	189.0	<20.0	30.3	40.0
Co	31.1	1.45	2.34	1.8
Fe(ppm)	120.0	0.73	3.39	0.3
Cr	14.0	14.5	81.8	<5.0

[a] Filby *et al.* (1978)
[b] Filby, Hunt and Khalil (1978).

(Filby *et al.*, 1980). Other potential problems identified in recent work include the formation of organometallic species and other undesirable metal moieties that may have serious effects on catalysts used in upgrading processes and the lack of Ti depletion in the SRC I process, compared with coal, due to the formation of organic complexes or formation of finely divided TiO_2 species.

3 Environmental Effects of Coal Utilization

3.1 Atmospheric Emissions of Trace Elements

3.1.1 Fallout in the vicinity of power plants

In recent years several field investigations have assessed the magnitude of atmospheric trace element deposition from power plants. Klein and Russell (1973) claimed that soils around a Michigan power plant were enriched in Cd, Co, Cr, Cu, Hg and Ni, with concentration trends reflected in the prevailing wind pattern. This contamination apparently extended 21 km from the plant and covered 113 km^2. However, despite the potential significance of these findings, Klein and Russell (1973) did not subject their data to statistical analysis.

Surface soils from industrial areas in Poland reportedly contained increased U and Th concentrations compared to rural regions; this difference was ascribed to the presence of power plants at the industrial sites (Jaworowski and Grzybowska, 1977). Similarly, Martin *et al.* (1971) found elevated air concentrations of Ra, Th and U 6 km downwind of a coal power plant. Another study in the U.S.A. (Styron, 1979) could, however, detect no increased radionuclide content in the vicinity of a power plant, possibly because of the use of an efficient particulate collection device in this facility.

An extensive ORNL survey, conducted around the Allen Steam Plant in Memphis, U.S.A. (Bolton *et al.*, 1973; Lyon, 1977) detected no impact of plant emissions on trace element levels in the surrounding environment. However, efforts to distinguish atmospheric deposition were hampered by the naturally elevated trace element content of soils in the Memphis area. Background soil concentrations of many trace elements were actually greater than those from the "enriched zone" of the Michigan survey (Klein and Russell, 1973). Additionally, U and Th levels in Memphis soils were four times greater than the values reported from industrial regions in Poland (Jaworowski and Grzybowska, 1977). Two elements of potential significance, As and Se, were unfortunately not examined fully in the Allen Plant study.

Efforts to detect a concentration trend in Hg with distance from the Allen Plant involved the analysis of moss samples but, in contrast to the Michigan study, were unsuccessful. Similarly, Crockett and Kinnison (1977) could find no accumulation of Hg in soils around the Four Corners plant, one of the largest power facilities in the U.S.A. The authors concluded that although emitted Hg may be initially deposited in the vicinity of the plant, its volatile nature leads to rapid vaporization. For this reason, they suggested the significance of Hg emissions should be evaluated on a regional rather than local basis. Aquatic systems adjacent to coal-fired power plants have also been examined for atmospheric Hg deposition (Anderson and Smith, 1977; Roffman, Kary and Hudgins, 1977). Analysis of sediments, water, macrophytes, fish and wildfowl revealed no significant elevation of Hg in these samples. All fish examined by Anderson and Smith (1977), for example, contained less than 0.5 $\mu g\ g^{-1}$ (wet weight), the FDA tolerance limit for consumption.

Serious environmental contamination from power plants is apparently confined to those facilities burning coal with unusually high trace element concentrations. Plants and soils downwind of a power plant in Nevada burning a high As coal ($>$500 $\mu g\ g^{-1}$) were

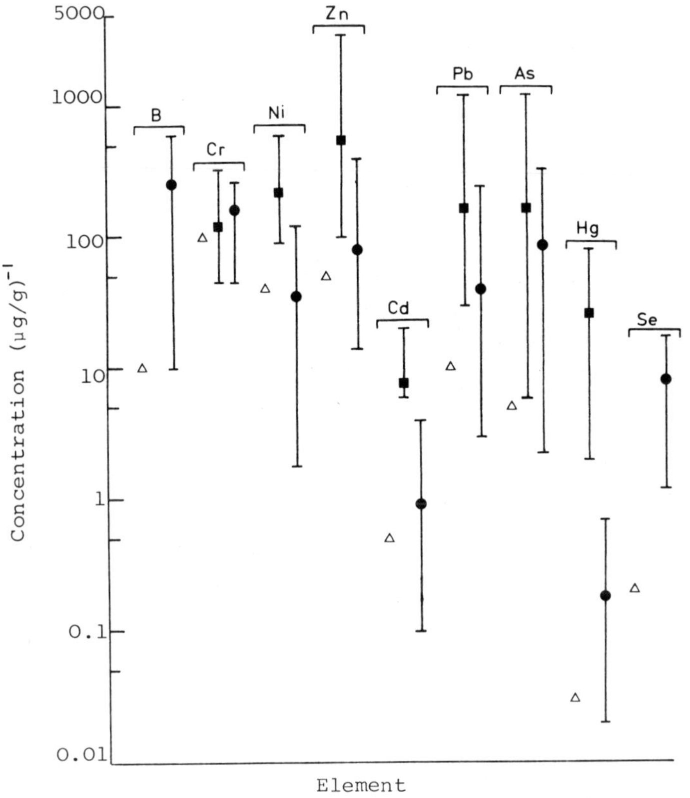

FIGURE 6.4 *Trace element concentrations (μg g^{-1}) in soil and in fly ashes from different regions of the U.S.A.: △, world average soil (Vinogradov, 1959; Bowen, 1966); ■, mean and range of 11 fly ash samples (Theis and Wirth, 1977); ●, mean and range of 23 fly ash samples (Furr et al., 1977).*

markedly enriched in As (Berry and Wallace, 1974). The authors stressed the hazard of As transfer to humans from the milk of those cows feeding in the area. The high As coal burned at a Czechoslovakian plant resulted not only in the contamination of adjacent pasture, but also the poisoning of livestock feeding in the area (Kuhnert, 1973).

The natural variability of trace element levels in soils requires, according to Wangen and Williams (1978), a 50% concentration increase to allow detection of any soil enrichment around power plants. Wangen and Williams (1978) further calculated that such increases would occur only when elemental fly ash concentrations were 100 times greater than local soils. In Figure 6.4, trace element levels in world average soil (Vinogradov, 1959; Bowen, 1966) are shown in relation to the range of values reported for fly ash. The ranges quoted were for fly ash either collected by electrostatic precipitator or ash which has passed through the precipitator, but still sampled within the stack. Fly ash concentrations in the upper range for As, Cd, Hg, Pb, Se and Zn are actually greater than soil by a factor of 100 or more. Presumably, the fly ash escaping from such plants would be further enriched in these elements. The degree of enrichment in such ashes would thus appear to be sufficient in theory to produce an observable impact in the vicinity of the power plant involved. However, if efficient precipitators or other means of particle attenuation are employed, then larger-sized particles are removed from the flue gases. If only very fine fly ash particles escape from the stack they will tend to be dispersed widely, although agglomeration of fine particles could increase effective particle size and density and result in increased deposition near the power station.

In conclusion, it is apparent that the impact on soils and biota of trace elements from power plants consuming coal of average trace element content is generally not detectable in relation to the natural elemental levels of these materials. Another factor responsible for the general absence of any detectable deposition in the vicinity of power plants is the dispersion of emissions over a wide area by the tall stacks employed at such facilities. The small particle association of several volatile elements would further enhance long-range dispersal, as would be the re-vaporization behaviour of Hg.

The source of enriched elements in urban aerosols

Several studies (Rahn, 1971, 1976; Gordon, Zoller and Gladney, 1973; Heindryckx, 1976) on the chemical composition of particles in urban aerosols revealed that lithophile elements, such as Fe, Mn, Sc and Th, have EFs close to 1, and are apparently derived from crustal weathering. Other, more volatile elements, including As, Br, Cd, Cu, Hg, I, Ni, Pb, Sb, Se and Zn are more enriched in urban particulates compared to crustal material. This finding was viewed with some concern, as many elements which showed the larger enrichments are also potentially toxic. As discussed earlier many volatile elements have large EFs on suspended fly ash particles. In the case of most elements, however, the EFs for urban aerosols are even

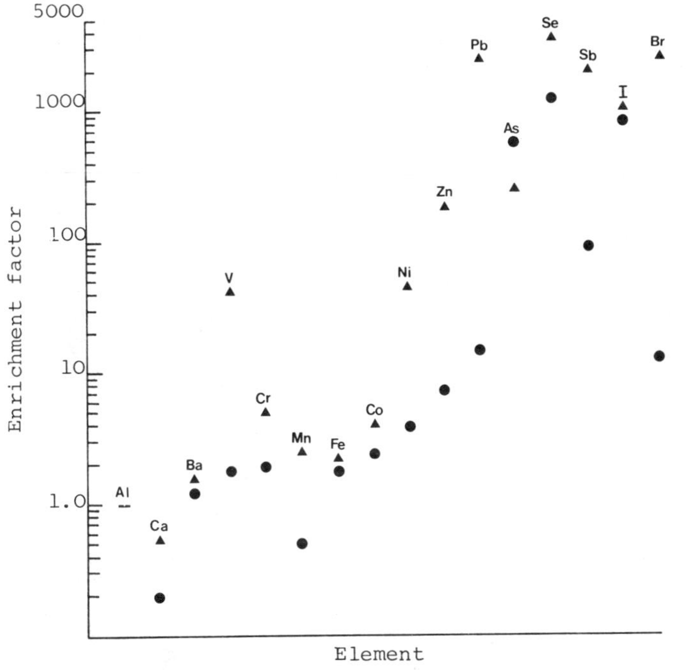

FIGURE 6.5 *Elemental enrichment factors relative to crustal abundance of suspended in-stack particles (●) at the Chalk Point plant and of ambient aerosols (▲) in the vicinity of the plant (Gladney, 1974)*

higher (Figure 6.5), suggesting that coal combustion is not responsible for their presence in urban particulates. Both As and Se are notable exceptions as these have similar EF values in suspended fly ash particles and urban aerosols (Gladney, 1974; Zoller *et al.*, 1974; Klein *et al.*, 1975). Direct sampling of a power plant plume revealed that the enrichment of As and Se persisted outside the power plant, while in the case of other elements, only the concentrations were enhanced compared to the background aerosol (Zoller *et al.*, 1974). Additionally, results of an attempt to identify air pollution sources by an elemental balance technique (Kowalczyk, Choquetter and Gordon, 1978) also indicated that coal combustion was the major source of only As and Se in the Washington DC aerosol.

Taken together, these findings suggest that beside As and Se the combustion of coal cannot account for the anomalous enrichment of trace elements in urban aerosols. However, examination of the

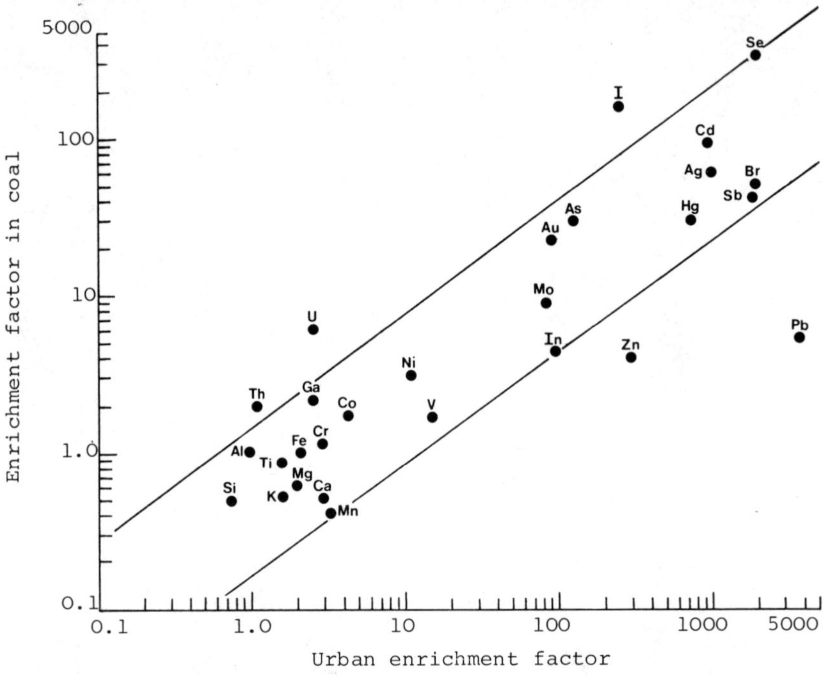

FIGURE 6.6 *Elemental enrichment factors in coal (with respect to crustal abundance) in relation to the corresponding enrichment factors of the urban aerosol (from Rahn, 1976)*

enrichment pattern in coal (Figure 6.6) reveals that those elements which display significant enrichment in coal are the same elements which are enriched in the urban aerosol. Moreover, the degree of elemental enrichment in coal is generally proportional to that in the aerosol, but is a factor of 10 lower. The additional enrichment caused by coal combustion is shown for Chalk Point power plant in Figure 6.7 (Gladney, 1974). For most elements, the combustion process contributes very little over and above the coal enrichment. Significant enhancement due to combustion itself only occurs for the more volatile elements, As, Hg, Pb, Sb and Se.

A possible reason for the close relationship between enrichment patterns in the urban atmosphere and coal is that coal combustion is actually responsible for the abundance of these elements in the aerosol. Certain elements will, of course, have specific sources; the large enrichment of Pb, for example, is probably associated with the

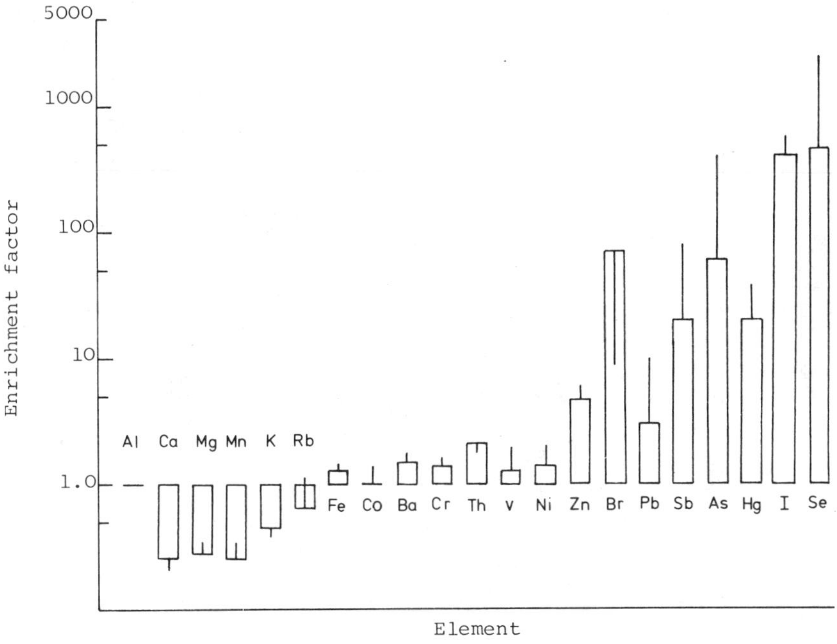

FIGURE 6.7 *Elemental enrichment factors with respect to crustal abundance at the Chalk Point power plant:* □, *coal;* |, *additional enrichment in fly ash attributable to combustion (Gladney, 1974, after Rahn, 1976)*

use of leaded petrol. An alternative explanation involves the enrichment of the urban aerosol by a variety of high-temperature process. If this is the case, then the parallel enrichment relationship between coal and urban aerosols (Figure 6.6) would require further explanation, involving an as yet unknown physico-chemical link between the enrichment of trace elements during coal formation and enrichment during high-temperature processes.

If coal combustion is the major source of certain enriched elements in urban aerosols, then account must be taken of the relatively low EF reported for many elements on suspended fly ash particles. This apparent discrepancy may be a reflection of the absence of reliable data on elemental EFs on sub-micron particles, as discussed earlier. Additionally, the chemical composition of aged power plant plumes may be very different from that of in-stack particulates. The greater settling velocities of larger fly ash particles would be expected to lead

TABLE 6.16 *Estimated global trace element emissions from coal combustion and other antropogenic sources in relation to natural sources*

	Global annual emission (× 10^9g)					
Source	As[a]	Cd[b]	Hg[c]	Ni[b]	Pb[b]	Zn[b]
Total natural	7.8	0.83	10	26	24.5	43.5
Coal combustion	5.8	0.06	0.54	0.66	14	15
Total anthropogenic	29	7.3	2.5	47	449	314

[a] Anthropogenic and natural source strengths from Walsh, Duce and Fasching (1979).
[b] All data from Nriagu (1979).
[c] Anthropogenic and natural source strengths from Miller, D. R. (unpublished).

to a fractionation with time, resulting in increased EFs for volatile elements, because of their small particle association. More information on the trace element composition and behaviour of power plant plumes with age is required before their contribution to the urban trace element load can be assessed.

Emissions in relation to other sources

The significance of atmospheric trace element emissions from coal combustion may be evaluated on a global scale by comparison with both the total natural and anthropogenic fluxes of the element in question. Relevant data for elements of toxic significance are shown in Table 6.16. The values reported have been obtained either directly from the literature or indirectly where information is lacking or considered inaccurate. Natural sources of importance include volcanoes, wind-blown dusts and vegetable exudates.

Anthropogenic emissions of Pb, Cd, Ni, and Zn are all far in excess of natural rates, but in each case coal combustion accounts for only a small proportion of the anthropogenic total. Nevertheless, the atmospheric emission of Pb from coal combustion occurs at over 50% of the rate of introduction by natural sources, its significance being masked by the consumption of leaded petrol (Nriagu, 1979).

Anthropogenic processes also dominate the introduction of As into the atmosphere, with the largest man-made contribution resulting from Cu production (Walsh, Duce and Fasching, 1979). In this example, the emission from coal combustion is also of importance, being comparable to the total natural rate. The figure estimated for coal combustion is, however, an order of magnitude greater than that calculated by Walsh, Duce and Fasching (1979) in a recent survey of atmospheric As emissions. The disparity arises because Walsh, Duce and Fasching (1979) employ an emission factor taken from the Allen Steam Plant study (Klein *et al.*, 1975) which is felt to be erroneously low. This is because no consideration is made of the large imbalance for As flow through the Allen Plant. In the present study, the atmospheric emission of As from coal combustion was obtained from the mean of two estimates. In the first, the average As content of coal was assumed to be 15 $\mu g\ g^{-1}$ (Table 6.5) and the annual global coal consumption 3×10^9 t (Quenon, 1980). An emission factor of 0.15 $g\ t^{-1}$ was employed, estimated from the wide range of reported values for the element (Natusch, Wallace and Evans, 1974; Schwitzgebel *et al.*, 1975; Roffman, Kary and Hudgins, 1977) giving an annual As emission of 6.8×10^3 t. The other method utilized the finding that the atmospheric release from nearly 700 power plants averaged 8% of the total ash in the coal burned (McBride *et al.*, 1978). Using an ash content of 13% in coal and an average As fly ash concentration of 150 $\mu g\ g^{-1}$ (Theis and Wirth, 1977) gives an annual emission of 4.7×10^3 t.

The total annual Hg emission from coal burning was estimated from an average Hg coal content of 0.2 $\mu g\ g^{-1}$ (Table 6.5) and an emission factor of 0.9 $g\ t^{-1}$ (Billings and Matson, 1972; Kalb, 1975) to give $0.54 \times 10^3\ g\ t^{-1}$. This figure is six times smaller than previous estimates by Joensuu (1971) or Heindryckx *et al.* (1974) who assumed an unrealistically high average Hg coal content and an emission factor of 1. However, the contribution from coal combustion is still significant, being over 20% of the total anthropogenic emission of the metal.

A number of comparisons of radioactive emissions from coal-fired power plants and nuclear power plants have been made (Eisenbud and Petrow, 1964, 1964; Hull, 1971; Martin *et al.*, 1971; Jordon and Schikarski, 1973; Jaworowski *et al.*, 1974; Martin, 1974; Lave, 1975; Illyin, Kniznikov and Barkhudov, 1977; Kniznikov and Barkhudov, 1977; McBride *et al.*, 1978). Some studies have emphasized the relatively greater biological significance of radionuclides released from coal combustion. Health impacts from coal-derived radionuclides will depend on a number of variables including the activity concentrations of radionuclides in the combusted coal (^{40}K, ^{238}U decay series and ^{232}Th decay series). Some of the worst prognoses have been based on the use of coal with a very high ^{238}U content but Beck *et al.* (1979) found that the activity concentration in coal samples varied over two orders of magnitude. The fly ash control measures adopted are also of importance as smaller, escaping fly ash particles may be considerably enriched with some decay series element. UNGA (1980) quotes fly ash atmospheric releases of between 1 and 30%.

Human health hazards

Particulate emissions from coal combustion may be inhaled by humans, and with stacks as tall as 200 m deposition may occur within a circle of 500 km radius (UNGA, 1980). As briefly discussed earlier, particle size is of major importance in determining the toxic significance of such emissions. This is because inhaled particles are deposited in different regions of the respiratory system depending on their effective aerodynamic particle size, as shown in Figure 6.8. Larger particles are preferentially deposited in the nasopharyngeal and tracheobronchial regions and are generally removed after a short time by cilial action and consequently swallowed. Elemental uptake from such particles will thus take place in the gastrointestinal system where absorption rates are much lower than in the lung (Norberg, 1976). Smaller particles are deposited more efficiently in the pulmonary region of the lung (Natusch, Wallace and Evans, 1974).

At present emissions from coal combustion have not been adequately characterized (section 2.1.1) but strong increases of several trace elements with decreasing particle size have been reported (Natusch, Wallace and Evans, 1974; Campbell *et al.*, 1978; Ondov, Ragaini and Biermann, 1978). For this reason it is again emphasized that the precise particle size distribution of toxic elements in suspended fly ash is required before the significance of such emissions can be evaluated. In addition, the possible surface predominance of volatile toxic elements (Natusch, Wallace and Evans, 1974) will result in intimate contact with the alveolar membrane and further enhance the toxicity of small particles. Finally, the chemical form of deposited elements in the lung requires evaluation as this will be an important factor in determining the solubility and hence uptake into the pulmonary bloodstream.

One aspect of human health concern is the possible carcinogenic properties of coal fly ash. Evidence in favour of this view derives from the reported mutagenicity of respirable fly ash filtrates to bacteria (Chrisp, Fisher and Lammert, 1978). (Mutagenic activity in bacterial test systems is regarded as a useful indicator of carcinogenicity in man.) Some of the mutagenic components present in very

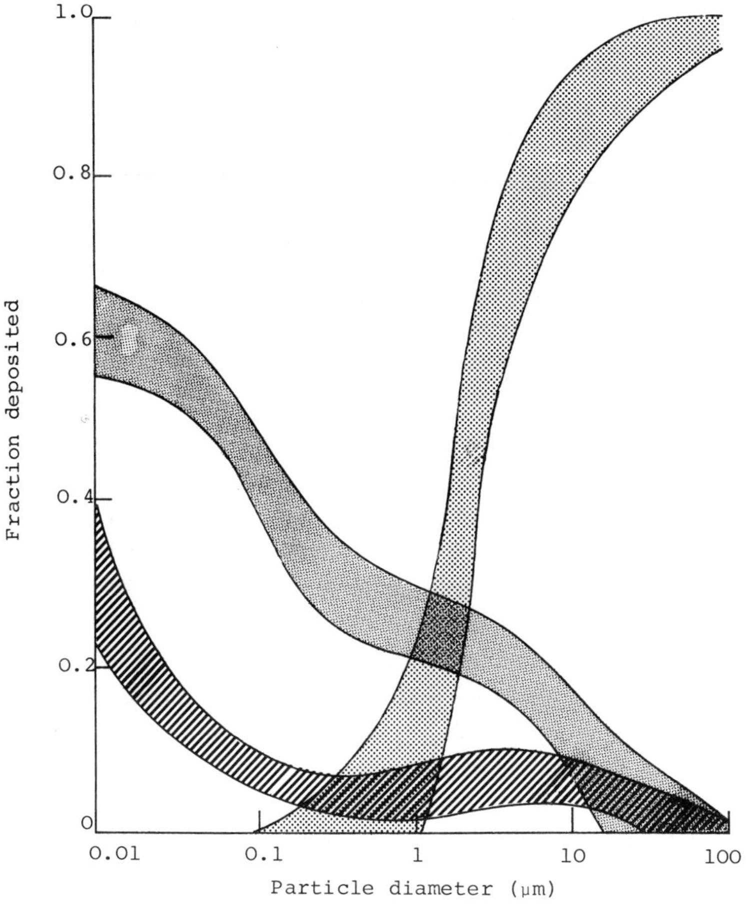

FIGURE 6.8 *Efficiency of particle deposition in the three respiratory tract compartments as a function of particle size:* ▦, *nasopharyngeal;* ▧, *tracheobronchial;* ▨, *pulmonary (Natusch, Wallace and Evans, 1974).*

fine fly ash were inorganic in nature (Fisher, Chrisp and Raabe, 1979). In addition, Chrisp, Fisher and Lammert (1978) noted that equine serum was more efficient in extracting the mutagens than saline, suggesting that such substances should be similarly soluble in alveolar lung fluid.

The three elements As, Cr, and Ni are present in high concentrations in fly ash particles and all possess carcinogenic forms of importance as As(III); Cr(VI) and Ni carbonyl. There is no information available on the abundance of the carcinogenic forms of As and Cr in fly ash, but Van Hook (1978) considers both are present. It is not thought that the thermolabile Ni carbonyl will be present in fly ash. Indentification of the possible carcinogenic substances in fine fly ash from coal combustion, and char and other residues from coal conversion processes, is required.

Arsenic contamination from the Czech power plant burning high As coal, described earlier (section 1.2.2), extended to humans in the area. Bencko and Symon (1977) reported that children residing up to 5 km from the plant contained elevated As levels in hair, urine and blood. Furthermore, hearing changes were reported in a group of boys living near the coal-fired power plant. Bencko and Symon (1977) tentatively assigned this to an As-induced effect, as there are known cases of auditory damage, including total deafness, caused by As poisoning. It was not clear whether the elevated As in these children was predominantly dietary in origin or resulted from respiratory uptake, as As air concentrations were extremely high near the plant, with a maximum of 69 $\mu g\ m^{-3}$ (Bencko *et al.*, 1976).

The radiological impacts of coal combustion

This has been assessed on a number of occasions (UNSCEAR, 1977; UNGA, 1979, 1980). Individuals will be exposed from several sources: suspensions of radioactive particles in the atmosphere, particles inhaled or ingested. Only the inhalation pathway was thought to be significant as a human health impact (Beck *et al.*, 1979) but more recent work is resulting in a revision of this view (see Section 2.2, Chapter 5). Estimates of collective dose commitments (UNGA, 1980) suggest that coal-fired power plant emissions are either equal to or less than those of a nuclear power plant in normal operation and that for both, the dose levels are extremely low, constituting only a negligible percentage of the dose from natural background radiation. A recent estimate of atmospheric radionuclide discharge from coal combustion is 56×10^{-3} GBq per MW(e)yr. This results in 3.7×10^{-3} man-Sv per MW(e)yr from inhalation, internal radiation and external irradiation due to the activity deposited, compared to an annual dose equivalent of 0.8×10^{6} man-Sv for the population of Europe and North America from natural irradiation.

One or two points need to be taken into consideration, however. The calculations for coal combustion are based on 1977 world coal production figures (3×10^{9} t). The lowest class of solubility in lungs is employed; bacterial mutagenicity studies (Chrisp, Fisher and Lammert, 1978) and aqueous extraction experiments (Theis and Wirth, 1977) indicate that at least some inorganic constituents will be soluble in the lung. Additionally the utilization of low-grade coals with elevated (20 $\mu g\ g^{-1}$) U content from parts of the western USA may pose significant health hazards on a local scale, not anticipated from the use of average activity concentrations.

3.2 Solid Wastes From Coal Utilization

3.2.1 Coal combustion

The solid wastes resulting from the combustion of coal in power plants are fly ash, bottom ash and slag, and if SO_2 scrubbers are employed, scrubber sludge. Fly ash is the most important waste material in modern pulverized coal power plants. Various forms of fly ash emission control devices are in use including mechanical or cyclone collectors, electrostatic precipitators, flue gas scrubbers and fabric filters. Ash retention may vary between 70 and 99% (UNGA, 1980). In the U.S.A. the EPA clean air standard to be eventually implemented corresponds to a release of about 1% of total ash. Problems of fly ash disposal primarily result from the enormous quantities collected; in the U.S.A. alone the figure was 60×10^6 t in 1976 (Scanlon and Duggan, 1979). Efforts have been made to employ fly ash for commercial purposes but stockpiles are increasing with time. Commercial utilization of fly ash is apparently more successful in the U.K. than the U.S.A., with about half the ash produced gainfully used in the U.K. (Brown, Ray and Ball, 1976), compared to 20% in the U.S.A. (Scanlon & Duggan, 1979). Disposal of fly ash is carried out by either wet storage in ash ponds or after removing water, in landfills. In both cases the disposed fly ash will be in close contact with adjacent soil and water systems.

Trace elements content of ash

Trace element concentrations in fly ash vary widely from one sample to another, reflecting differences in the original coal composition as well as boiler-operating conditions. Reported values from two studies of fly ash from the U.S.A. are shown in Figure 6.4. The investigation by Furr *et al.* (1977) is thought to provide representative values as levels were measured in 23 samples from 21 states in the U.S.A.

Mobilization of trace elements from ash ponds

The possible contamination of water and soils by the leaching of trace elements from fly ash disposal ponds has only recently received attention. Theis and his co-workers (Theis *et al.*, 1978; Theis and Richter, 1979) reported that although trace element concentrations in an ash pond were elevated, levels rapidly decrease in ground-water and soil with distance from the pond. In accord with experimental extraction studies (Theis and Wirth, 1977), Theis *et al.* (1978) noted that the partitioning of trace elements between water and soil, as well as their attenuation around the ash pond, was most affected by the pH and the oxides of manganese and iron. Accumulation of metals in the adjacent soil was primarily due to a complex interaction between precipitation and absorption on to these metal oxides. It was pointed out by Theis *et al.* (1978) that unless these metal oxides were replenished, the absorptive capacity of the system was likely to be exhausted, resulting in a freer migration of trace elements from the ash pond. In view of its large sorptive capacity, manganese oxide additions to ash ponds may well be a useful control mechanism to reduce leaching of trace elements from such sites.

The pH of ash pond leachate also exerts an important influence upon trace element behaviour. The alkaline nature of most fly ashes (Townsend and Gilham, 1975) results in the precipitation of insoluble hydroxides of many toxic metals in the ash pond itself. Elements with anionic character such as As, B, Cr, F and Se are more soluble in the alkaline environment and significantly elevated concentrations of these elements have been reported in ash ponds (Theis *et al.*, 1978) and their effluent water (Dreeson *et al.*, 1977). Future investigations of contamination from ash ponds into effluent waters should thus focus upon the behaviour of these elements. It should be noted, however, that attentuation of these elements in soil adjacent to the ash pond still occurs and they are probably absorbed on to positive iron hydroxide (Theis and Richter, 1979).

3.2.2 Coal conversion processes

The range of solid wastes derived from coal conversion processes not only include a bottom ash and a fly ash, analogous to the products of conventional coal combustion, but also a carbonaceous char or residue. Char from gasification may be combusted to provide energy, while liquefaction residues may undergo gasification to produce synthetic gas. Thus, a further series of solid inorganic wastes will be derived from such processes. Additionally, it is possible that char or residue will be produced in excess of demand, necessitating the disposal of these wastes. The intended combustion of synthetic liquid fuels in power plants will also produce significant quantities of ash. Such produce residues will consist of a bottom ash and, if collection devices are employed, a fly ash. Finally, spent catalysts from conversion processes which cannot be reclaimed will also require disposal.

The utilization of inorganic ashes from coal conversion, for such purposes as road construction, may occur on a commercial scale, but this is unlikely to solve the disposal problem of these wastes. Disposal is expected to take place in landfills or ash ponds adjacent to the conversion facility. It is probable that the spent catalysts and any unwanted char or residue will also be disposed of in this manner. In the case of inorganic ashes, it is expected that the mobilization and bioaccumulation of trace elements from such disposal areas will be similar to that found for fly ash derived from conventional coal combustion. Certain S compounds, absent from solid waste from combustion processes, may still be present in conversion process wastes. Attention should be paid also to the behaviour of trace elements in the fly ash produced by the Koppers-Totzek process. This is because the operating conditions of this process indicate that concentrations of several potentially toxic elements may actually be greater than found for coal combustion. The mobilization potential of trace elements present in the waste organic residues has not been investigated. It is thought unlikely that this will differ substantially from the limited migration of trace elements observed in the vicinity of fly ash disposal sites (Theis *et al.*, 1978). Contamination of water and soils by the Co, Cr, Fe, Mo and V present in spent catalysts is expected to occur adjacent to the disposal site, but the likely extent of such contamination is unknown and requires investigation.

4 Conclusions

4.1 Summary of Evidence from Existing Combustion Processes

The environmental and human health implications of trace element release from coal utilization are of current concern, notably because the quantities of coal combusted are at present very large and are planned to further increase in the near future (U.S. National Committee for Geochemistry, 1980).

The results of investigations into the combustion behaviour of trace elements in power plants can be explained by the physico-chemical properties of the elements and their compounds, and by the operating conditions within the power plant. Such studies have enabled the classification of trace elements according to the manner in which they are partitioned between the various exit streams and by their particle size distribution in suspended fly ash. These findings have been rationalized in terms of a volatilization-condensation mechanism, involving the vaporization of volatile elements during combustion and their subsequent condensation on fly ash particles. This model predicts that there will be dramatic increases in the concentration of volatilized elements on sub-micron particles and these elements will be preferentially enriched on the surface of such particles. These points are of significance as particles of this size are retained least efficiently by collection devices, have the longest atmospheric residence times and are deposited most efficiently in the pulmonary region of the respiratory system. Several recent investigations have, however, reported smaller enrichments of these elements than predicted by this model, leading to the development of a more complex mechanism to account for the observed behaviour of trace elements in power plants. Whatever the precise mechanism, it is essential that the extent of enrichment of several potentially toxic elements of sub-micron particles be clarified. More information on the surface predominance of volatile elements and their chemical form in fly ash is also required, as both factors are important in determining the toxicity and solubility of these elements. Such studies should extend to emissions from domestic heating appliances, as evidence from one investigation indicates that certain elements may also be highly enriched in the suspended particles from these units.

The full characterization of emissions from coal-fired power plants, together with an understanding of the effects of ageing on the composition of such plumes, will allow an assessment of the importance of coal combustion in contributing to the large enrichment of trace elements in urban aerosols. In particular, the significance of the high enrichment factors reported for As and Se in power plan emissions requires substantiation.

Evaluation of the potential human health hazards from coal combustion also requires the accurate characterization of trace element emissions. In the context of human health, the possible carcinogenic properties of fly ash also require further investigation, with special attention paid to the presence of the carcinogenic forms of As and Cr in respirable-sized fly ash particles.

Only in one case of As contamination has serious environmental pollution been directly linked to trace element emissions from coal combustion. The power plant responsible consumed a low-grade coal with elevated As content, and children in the vicinity of the facility exhibited symptoms associated with As toxicity. It is not known whether other power plants in the region also consume coal with high As concentrations or if the use of improved particulate control technology would significantly reduce As emissions. It is suggested that the screening of coals suspected to contain high As concentrations be carried out, and the use of such coals be avoided, especially in facilities without adequate control devices.

Estimates of the radiological impact to humans from coal combustion indicate that this source represents a very small fraction of the collective dose. However, the utilization of uraniferous coals from the western U.S.A., or from other parts of the world, would be expected to substantially increase the radiological exposure to humans residing in the vicinity of such plants.

The importance of coal combustion as a source of atmospheric trace elements has been assessed on a global scale. Emissions of As from coal combustion apparently rival the total natural emission rate for this element. Although the other elements considered in this report are released in significant amounts, the contribution to the total anthropogenic flux appears to be relatively small, with the exception of Hg. Earlier estimates of the global releases of Hg from coal combustion are thought to be too large, as unrealistically high Hg values were assumed. Such errors highlight the necessity for both accurate and representative trace element values in coal, to enable the significance of this source of contamination to be determined.

The mobilization of trace elements from ash disposal areas has only recently received attention. The limited data available indicate there is rapid attenuation in soils and ground-water close to the disposal site. Contamination of adjacent water courses by direct run-off from disposal ponds results in decreased water quality and the accumulation of trace elements toxic to aquatic organisms. Present information suggests that such contamination is also restricted to the close vicinity of the disposal ponds. However, the expected increase in coal utilization will result in greater quantities of fly ash requiring disposal, which will place a greater strain on the absorptive capacities of soils. For this reason, control techniques, such as the addition of manganese oxide, may be needed to reduce the leaching of trace elements from ash disposal sites.

In conclusion, it is clear that at present there is insufficient information to allow an adequate evaluation of the environmental and health impacts of trace element release from current and projected coal utilization. From the evidence available it would appear that coal combustion will make a contribution, along with other anthropogenic sources, to the trace element load in the atmosphere. The significance of this contribution will be dependent upon the ambient air quality, the proximity of other anthropogenic sources and the concentrations of volatile trace elements in the coal combusted. In the case of solid wastes, the limited information available indicates that trace elements are efficiently retained within the fly ash disposal sites, but long-term studies are needed to determine leaching rates upon weathering. For this reason, it is felt unrealistic to compare the quantities of trace elements discharged in disposal sites with the natural weathering rates of these elements. However, the mobilization and bioaccumulation of trace elements in fly ash used to ameliorate agricultural soils requires further investigation.

4.2 Predicting Likely Effects of New Conversion Technologies

The assessment of trace element contamination from future coal conversion technologies was based upon information from pre-commercial conversion units, recent published material on commercial gasification and liquefaction plants, and extrapolation from studies of conventional coal-fired power facilities. In coal-fired power plants the operating conditions are known to be important in determining the partitioning of elements between the various exit streams. Unfortunately, the large differences in operating conditions between the various processes still being developed for coal conversion make predictions for future trace element impact difficult. Nevertheless, most of the conversion processes discussed in this section employ high pressures and lower temperatures than conventional coal-fired power plants. These conditions are expected to reduce the vaporization of volatile elements, resulting in a less marked partitioning between fly ash and bottom ash and a smaller enrichment of such elements on sub-micron particles. Thus, atmospheric emissions from coal conversion facilities are likely to exert less impact on the loading of ambient aerosols than coal combustion.

Another important factor affecting the environmental significance of coal conversion is the trace element content of the feed coal. Preference may be given for the selection of coals which exhibit high hydrogenation rates. This may result in the use of coals which contain a high mineral content and elevated concentrations of certain trace elements.

From the limited information available it would appear that gasifier char retains a large proportion of the trace elements present in the feed coal. The proposed combustion of char in utility boilers may thus result in the emission of fly ash enriched in volatile trace elements.

The use of catalysts in coal conversion processes adds another dimension to the environmental implications of this technology which requires assessment. Spent catalysts which are not reclaimed will pose a disposal problem, the importance depending on the amounts of catalysts used. The metals present in these catalyts may contaminate the conversion products as well as the effluent streams.

The trace elements retained in the conversion products will also be mobilized when those products are combusted. From the limited data it would appear that liquefaction products contain trace element concentrations in the range found for coal, or with significant reductions in certain elements (Table 6.12). The trace element composition of emissions from the burning of these oils may be no greater than that found for coal combustion. Gasification products undergo cleaning procedures which appear to remove most trace elements. However, some available evidence indicates that Hg and F may still be present in the final gas product (Oldham and Wetherold, 1977). If this is the case, then further cleaning may be required in order to remove these toxic elements.

The large range of inorganic wastes from coal conversion processes are derived from the coal conversion process itself, from the further combustion of char and gasification of residue, and from the combustion of liquefaction products. In addition, spent catalysts and unwanted chars and residues may also require disposal. In most cases it is expected that the behaviour of trace elements in sites selected for disposal will be similar to that reported for the fly ash disposal ponds associated with coal combustion. However, attention should be paid to the extent of leaching of trace elements from such sites, especially where spent catalysts have simultaneously been disposed.

Experimental, theoretical and limited commercial experience of coal conversion processes indicate that a number of environmental problems may arise as the conversion processes are developed. As far as gasification is concerned, water treatment processes for condensate water containing high concentrations of Hg, As, Sb, Pb and other metals, disposal of water treatment solids containing high toxic metal concentrations, the combustion of char material, disposal of bottom ashes and the disposal of spent catalysts and metal deposits are all areas of environmental concern that need consideration and research. Development of liquefaction processes will engender concern over elements with volatile behaviour such as Hg, Sb, and As, particularly their presence in process waters and some light oils. Those treatment processes that have been shown to effectively reduce toxic metals in effluent waters produce biotreatment sludges containing high concentrations of Hg, Se, As and other elements. These sludges, and deposits from fouled process units, may produce significant disposal problems.

5 References

ABERNATHY, R. F., PETERSON, M.I. and GIBSON, F. H. (1969). Spetrochemical analyses of coal ash for trace elements. U.S. Bureau of Mines Report Investigation No. 7281.

ANDERSON, W. L. and SMITH, K. E. (1977). Dynamics of mercury at a coal-fired power plant and adjacent cooling lake. *Environmental Science and Technology*, **11**, 75–80.

ANDERSON, G. L., HILL, A. H. and FLEMING, K. K. (1979). *Symposium Proceedings on Environmental Aspects of Fuel Conversion Technology, IV*, Hollywood, U.S. EPA 600/7–79–217, p. 303.

ANDREN, A. W., KLEIN, D. H. and TALMI, Y. (1975). Selenium in coal-fired steam plant emissions. *Environmental Science and Technology*, **9**, 856–8.

BARBER, D. E. and GIORGIO, H. R. (1977). Gamma-ray activity in bituminous, sub-bituminous and lignite coals. *Health Physics*, **32**, 83–8.

BECK, H. L., GOGOLAK, C. V. and MILLER, K. M. (1979). Perturbations on the utilization of coal as an energy source. *Proceedings of the Natural Radiation Environment, III.*

BENCKO, V. and SYMON, K. (1977). Test of environmental exposure to arsenic and hearing changes in exposed children. *Environmental Health Perspectives*, **19**, 95–101.

BENCKO, V., SYMON, K., CHLADEK, V. and PIHRT, J. (1976). Health aspects of burning coal with a high arsenic content. II. Hearing changes in exposed children. *Environmental Research*, **13**, 386–95.

BERRY, W. L. and WALLACE, A. (1974). Trace elements in the environment: their role and potential toxicity as related to fossil fuels. A preliminary study. University of California, Los Angeles, Report No. UCLA–12–946.

BERTINE, K. K. and GOLDBERG, E. D. (1971). Fossil fuel combustion and the major sedimentary cycle. *Science*, **173**, 233–5.

BETHELL, F. V. (1962). The distribution and origin of minor elements in coal. *The British Coal Utilisation Research Association*, **25**, 401–30.

BILLINGS, C. E. and MATSON, W. R. (1972). Mercury emissions from coal combustion. *Science*, **176**, 1232–3.

BLOCK, C. and DAMS, R. (1975). Inorganic composition of Belgian coal and coal ashes. *Environmental Science and Technology*, **9**, 146–50.

BLOCK, C. and DAMS, R. (1976). Study of fly ash emission during combustion of coal. *Environmental Science and Technology*, **10**, 1011–17.

Bolton, N. E., Van Hook, R. I., Fulkerson, W., Lyon, W. S., Andren, A. E. and Emery, J. F. (1973). Trace element measurements at the coal-fired Allen Steam Plant. Progress Report, June 1971–January 1973. Report No. ORNL–NSF–EPL43. Oak Ridge National Laboratory Tennessee.

Bombaugh, K. J., Corbett, W. E., Lee, K. W. and Seames, W. S. (1980). An environmentally based evaluation of the multi-media discharge from the Lurgi coal gasification system at Kosovo. *Symposium on Environmental Aspects of Fuel Conversion Technology, V.* St Louis, Missouri, U.S. Environmental Protection Agency.

Bowen, H. J. M. (1966). *Trace Elements in Biochemistry.* Academic Press, New York.

Brown, J., Ray, N. J. and Ball, M. (1976). The disposal of pulverised fuel ash in water supply catchment areas. *Water Research*, **10**, 1115–21.

Cahill, R. A. (1974). A Study of the Trace Element Distribution in Petroleum. M.S. Thesis, University of Maryland.

Campbell, J. A., Laul, J. C., Nielsen, K. K. and Smith, R. D. (1978). Separation and chemical characterization of finely-sized particles. *Analytical Chemistry*, **50**, 1032–40.

Campbell, J. A., Smith, R. D., Davis, L. E. and Smith, K. L. (1979). Characterization of micron-sized fly-ash particles by X-ray photoelectron spectroscopy (ESCA). *The Science of the Total Environment*, **12**, 75–85.

Chatterjee, P. K. and Pooley, F. D. (1977). An examination of some trace elements in South Wales coals. *Proceedings of the Australian Institute of Mining and Metallurgy*, **263**, 19–30.

Chrisp, C. E., Fisher, G. L. and Lammert, J. E. (1978). Mutagenicity of filtrates from respirable coal fly ash. *Science*, **199**, 73–5.

Clark, M. C. and Swaine, D. J. (1962). *Trace Elements in Coal.* CSIRO Technical Communication No. 45.

Coles, D. G., Ragaini, R. C. and Ondov, J. M. (1978). Behaviour of natural radionuclides in western coal-fired power plants. *Environmental Science and Technology*, **12**, 442–6.

Coles, D. G., Ragaini, R. C., Ondov, J. M., Fisher, G. L., Silberman, D. and Prentice, B. A. (1979). Chemical studies of stack fly ash from a coal-fired power plant. *Environmental Science and Technology*, **13**, 455–9.

Craun, J. C. and Massey, M. J. (1979). *Proceedings of the Symposium on Potential Health and Environmental Effects of Synthetic Fossil Fuel Technologies,* U.S. DOE CONF 780903, p. 42.

Crockett, A. B. and Kinnison, R. R. (1977). Mercury distribution in soils around a large coal-fired power plant. U.S. Environmental Protection Agency Report, 600/3–77–063.

Denson, N. M. (1959). Uranium in coal in the western United States. *United States Geological Survey Bulletin*, No. 1055A. United States Geological Survey, Washington, DC.

Dreeson, D. R., Gladney, E. S., Owens, J. W., Perkins, B. L., Wienke, C. L. and Wangen, L. E. (1977). Comparison of levels of trace elements extracted from fly ash and levels found in effluent waters from a coal-fired power plant. *Environmental Science and Technology*, **11**, 1017–21.

Dvorak, A. J. (1977). The environmental effects of using coal for generating electricity. U.S. Nuclear Regulatory Commission Report No. NUREG–0252. Washington, DC.

EISENBUD, M. and PETROW, H. G. (1964). Radioactivity in the atmosphere effluents of power plants that use fossil fuels. *Science*, **144**, 288–9.

FILBY, R. H., HUNT, M. L. and KAHLIL, S. R. (1978). *Interim Report on Trace Elements in the SRC I Process.* U.S. Department of Energy.

FILBY, R. H., SHAH, K. R., HUNT, M. L., KHALIL, S. R. and SAUTTER, C. A. (1978). *Solvent Refined Coal (SRC) Processes. Trace Elements.* Research & Development Report No. 53, vol. III, Part 6. U.S. Department of Energy Report FE/496–T–17.

FILBY, R. H., KHALIL, S. R., GRIMM, C. A., EKABARUM, V. and HUNT, L. M. (1980). Interim Report on Trace Elements in the SRC II Process U.S. Department of Energy Report.

FIRTH, J. N. M. (1971). The Mineralogy of the South Wales Coalfield. Ph.D. thesis, University of Bristol.

FISHER, G. L., CHRISP, C. E. and RAABE, O. G. (1979). Physical factors affecting the mutagenicity of fly ash from a coal-fired power plant. *Science*, **204**, 879–81.

FORNEY, A. J. (1974). Analyses of tars, chars, gases and water found in effluents from the Synthane Process. *Technical Progress Report* No. 76, Pittsburgh Energy Research Centre, Pittsburgh, Pa.

FURR, A. K., PARKINSON, T. F., HINRICHS, R. A., VAN CAMPEN, D. R., BACHE, C. A., GUTENMANN, W. H., ST JOHN, L. E., PAKKALA, I. S. and LISK, D. J. (1977). National survey of elements and radioactivity in fly ash soil mixtures. *Environmental Science and Technology,* **11**, 1194–201.

GHASSEMI, M., CRAWFORD, K. and QUINLIVAN, S. (1979). *Environmental Assessment Report: Lurgi Coal Gasification Systems for SNG.* U.S. Environmental Protection Agency Report EPA–600/7–79–120 Washington, DC.

GIBSON, J. (1979). *Coal and the Environment.* National Coal Board, London.

GLADNEY, E. S., SMALL, J. A., GORDON, G. E. and ZOLLER, W. H. (1976). Composition and size-distribution of in-stack particulate material at a coal-fired power plant. *Atmospheric Environment*, **10**, 1071–7.

GORDON, G. E., ZOLLER, W. H. and GLADNEY, E. S. (1973). Abnormally enriched trace elements in the atmosphere. *Trace Substances in Environmental Health,* VII (ed. by D. D. Hemphill), pp. 167–74. University of Missouri, Columbia, Missouri.

HEINDRYCKX, R. (1976). Comparison of the mass-size functions of elements in the aerosols of the Gent industrial district with data from other areas. Some physico-chemical implications. *Atmospheric Environment*, **10**, 65–71.

HEINDRYCKX, R., DEMUNYNCK, M., DAMS, R., JANSSENS, M. and RAHN, K. A. (1974). Mercury and cadmium in Belgian aerosols. *Problems of the Contamination of Man and his Environment*, pp. 135–48. CEC, Luxembourg.

HEINRICHS, H. (1977). Emissions of 22 elements from brown coal combustion. *Naturwissenschaften*, **64**, 479–81.

HULL, A. P. (1971). Radiation in perspective: some comparisons of the environmental risks from nuclear and fossil fueled power plants. *Nuclear Safety*, **12**, 185–96.

ILLYIN, L. A., KNIZNIKOV, V. A. and BARKHUDOV, R. M. (1977). A relative risk estimation of excessive frequency of malignant tumours in populations due to discharges into the atmosphere from fossil-fuel and nuclear power stations. In *IV International Congress*, Paris, pp. 189–93.

JAWOROWSKI, Z. and GRZYBOWSKA, D. (1977). Natural radionuclides in industrial and rural soils. *The Science of the Total Environment*, **7**, 45–52.

JOENSUU, O. I. (1971). Fossil fuels as a source of mercury pollution. *Science*, **172**, 1027–8.

JORDON, S. and SCHIKARSKI, W. (1973). Evaluation of radioactive and non-radioactive trace constituents emitted from fossil-fuel and nuclear power plants. *Environmental Behaviour of Radionuclides Released in the Nuclear Industry*, pp. 525–35. IAEA Publication STI/PUB/345, Vienna.

KAAKINEN, J. W., JORDEN, R. M., LAWASANI, M. H. and WEST, R. E. (1975). Trace element behaviour in coal-fired power plants. *Environmental Science and Technology*, **9**, 862–9.

KALB, G. W. (1975). Total mercury mass balance at a coal-fired power plant. In *Trace Elements in Fuel* (ed. by S.P. Babu), pp. 154–74. Symposium of the 166th Meeting of the American Chemical Society, 1973. American Chemical Society, Washington, DC.

KLEIN, D. H. and RUSSELL, P. (1973). Heavy metals: fallout around a power plant. *Environmental Science and Technology*, **7**, 357–8.

KLEIN, D. H., ANDREN, A. W., CARTER, J. A., EMERY, J. F., FELDMAN, C., FULKERSON, W., LYON, W. S., OGLE, J. C., TALMI, Y., VAN HOOK, R. I. and BOLTON, N. (1975). Pathways of thirty seven trace elements through coal-fired power plant. *Environmental Science and Technology*, **9**, 937–9.

KNIZNIKOV, V. A. and BARKHUDOV, R. M. (1977). Comparative assessment of radiation danger to populations from TES and APS atmospheric emissions. *Atomic Energy*, no. 3, vol. 43.

KOPPENHAAL, D. W., SCHULTZ, H., LETT, R. G., BROWN, F. R., BOOHER, H. B. and HATTMAN, E. A. (1979). Preprints ASC Div. *Fuel Chemistry*, **24**, 299.

KOWALCZYK, G. S., CHOQUETTER, C. E. and GORDON, G. E. (1978). Chemical element balances and identification of air pollution sources in Washington, DC. *Atmospheric Environment*, **12**, 1143–53.

KUHNERT, M. (1973). Emissions and their importance for cattle breeding and human health. *Zeitschrift für die Gesamte Hygiene und Ihre Grenzgebiete*, **19**, 89–98.

LAVE, L. B. (1975). Health effects of electricity generation from coal, oil and nuclear fuel. *Symposium Proceedings on Energy and the Environment: A Risk Benefit Approach*, pp. 63–85. Birmingham.

LETT, R. G., SCHMIDT, C. E., DE SANTIS, R. R. and SHARKEY, A. G. JR. (1977) PERC/RI12, Pittsburgh Energy Technology Centre.

LINTON, R. W., LOH, A., NATUSCH, D. F. S., EVANS, C. A. and WILLIAMS, P. (1976). Surface predominance of trace elements in airborne particles. *Science*, **191**, 852–4.

LYON W. S. (1977). *Trace Element Measurements at the Allen Coal-Fired Steam Plant*. CRC Press, Cleveland, Ohio.

MCBRIDE, J. P., MOORE, R. E., WITHERSPOON, J. P. and BLANCO, R. E. (1978). Radiological impact of airborne effluents of coal-fired and nuclear power plants. *Science*, **202**, 1045–50.

MARTIN, Y. E. (1974). Comparative population radiation dose commitments of nuclear and fossil-fuel electric power cycles. *Proceedings of Symposium on Population Exposures*, p. 137, Health Physics Society, Knoxville, Tennessee.

MARTIN, J. E., HARWARD, E. D., OAKLEY, D. T., SMITH, J. M. and BEDROSIAN, P. H. (1971). Radioactivity from fossil fuel and nuclear power plants. *Proceedings of Symposium on Environmental Aspects of Nuclear Power Stations*, pp. 352–57. International Atomic Energy Agency, Vienna.

MASON, B. (1966). *Principles of Geochemistry*. 3rd. edn. Wiley, New York.

MUKHERJEE, D. K. and CHOWDHURY, P. B. (1976). Catalytic effect of mineral matter in a North Assam coal on hydrogenation. *Fuel*, **55**, 4–8.

NATUSCH, D. F. S., WALLACE, J. R. and EVANS, C. A. (1974). Toxic trace elements:

preferential concentration on respirable particles. *Science*, **183**, 202–4.

NORBERG, G. F. (ed.) (1976). *Effects and Dose Response Relationships of Toxic Metals.* Elsevier Scientific Publishing, Amsterdam.

NRIAGU, J. O. (1979). Global inventory of natural and anthropogenic emissions of trace metals to the atmosphere. *Nature*, **279**, 409–10.

OLDHAM, R. G. and WETHEROLD, R. G. (1977). Assessment, selection, and development of procedures for determining the environmental acceptability of synthetic fuel plants. National Technical Information Service. Report No. FE 1795–3.

ONDOV, J. M., ZOLLER, W. H., OLMEZ, K., ARAS, N. K., GORDON, G. E., RANCITELLI, L. A., ABLE, K. H., FILBY, R. H., SHAH, K. R. and RAGAINI, R. C. (1975). Elemental concentrations in the National Bureau of Standards' environmental coal and fly ash standard of reference materials. *Analytical Chemistry*, **47**, 1102–9.

ONDOV, J. M., RAGAINI, R. C. and BIERMANN, A. H.(1978). Elemental particle size emissions from coal-fired power plants: use of an inertial cascade impactor. *Atmospheric Environment*, **12**, 1175–85.

PATTERSON, R. K. (1980). Ambient air downwind of the Kosovo gasification complex: *Compendium Symposium on Environmental Aspects of Fuel Conversion Technology, V.* St. Louis, Missouri, U.S. Environmental Protection Agency.

PITTSBURG & MIDWAY COAL MINING CO. (1978). Solvent Refined Coal (SRC) Processes: Monthly Report for March FE/496–148, Pittsburg & Midway Coal Mining Company, Merrian, Kansas.

QUENON, R. H. (1980). World coal trade through the year 2000–rapid growth and political challenges. In *11th World Energy Conference, Munich*, vol. 1A, pp. 383–402.

RAHN, K. A. (1971). Sources of Trace Elements in Aerosols–an Approach to Clean Air. Ph.D. thesis, University of Michigan.

RAHN, K. A. (1976). *The Chemical Composition of the Atmospheric Aerosol.* Technical Report. Graduate School of Oceanography, University of Rhode Island.

RICKETTS, T. S. (1964). Modern methods of gas manufacture including the Lurgi process. *Journal of the Institute of Fuel*, **37**, 382–41.

ROFFMAN, H. K., KARY, R. E. and HUDGINS, T. (1977). Ecological distribution of trace elements emitted by coal-burning power generating units employing scrubbers and electrostatic precipitators. *Fourth Symposium on Coal Utilization*, pp. 192–215. Louisville, Kentucky.

RUCH, R. R., GLUSKOTER, H. J. and SHIMP, N. F. (1974). Occurrence and distribution of potentially volatile trace elements in coal. Final report. *Environmental Geology Notes,* no. 72. Illinois Geological Survey, Urbana, Illinois.

SCANLON, D. H. and DUGGAN, J. C. (1979). Growth and element uptake of woody plants on fly ash. *Environmental Science and Technology*, **13**, 311–15.

SCHULTZ, H., GIBBON, G. A., HATTMAN, E. A., BOOHER, H. B. and ADKINS, J. W. (1977). PERC Report, PERC/RI–77/2.

SCHULTZ, H., GIBBON, G. A., HATTMAN, E. A., BOOHER, H. B. and ADKINS, J. W. (1978). *Analytical Methods for Coal and Coal Products*, (ed. by C. Karr Jr.), chap. 19, vol. 1. Academic Press, New York.

SCHWITZGEBEL, K., MESEROLE, F. B., OLDHAM, R. G., MAGEE, R. A., MESICH, F.

G. and Thoem, T. L. (1975). Trace elements discharge from coal-fired power plants. *International Conference on Heavy Metals in the Environment*, pp. 533–51. Proceedings of the Toronto Conference, Ontario, Canada.

Smith, K. R. (1977). The coal-uranium breeder: uranium from coal. *Energy*, **2**, 171–7.

Smith, R. D., Campbell, J. A. and Nielson, K. K. (1979). Characterization and formation of submicron particles in coal-fired plants. *Atmospheric Environment*, **13**, 607–17.

Styron, C. E. (1979). An assessment of natural radionuclides in the coal fuel cycle. *Proceedings of the Natural Radiation Environment*, III.

Swaine, D. J. (1977). Trace elements in coal. *Trace Substances in Environmental Health*, XI (ed. by D. D. Hemphill), pp. 107-16. University of Missouri, Columbia, Missouri.

Swaine, D. J. (1979). Trace elements in Australian bituminous coals and fly ashes. *Combustion of Pulverized Coal - The Effect of Mineral Matter* (ed. by I. M. Stewart and T. F. Wall), pp. W3, 14–18. University of Newcastle, N.S.W.

Swanson, V. E., Medlin, J. H., Hatch, J. R., Colemen, S. L., Wood, G. H., Woodruff, S. D. and Hildebrand, R. T. (1976). Collection, chemical analysis and evaluation of coal samples in 1975. U.S. Department of Interior, Geological Survey. Open file report 76–468.

Theis, T. L. and Richter, R. O. (1979). Chemical speciation of heavy metals in power plants ash pond leachate. *Environmental Science and Technology*, **13**, 219–24.

Theis, T. L. and Wirth, J. L. (1977). Sorptive behaviour of trace metals on fly ash in aqueous systems. *Environmental Science and Technology*, **11**, 1096–1100.

Theis, T. L., Westrick, J. D., Hsu, C. K. and Marley, J. J. (1978). Field investigations of trace metals in ground water from fly ash disposal. *Journal of the Water Pollution Control Federation*, **50**, 2457–69.

Thomas, W. C., Trede, K. N. and Page, G. C. (1979). *Environmental Assessment: Source Test and Evaluation Report - Wellman-Galusha (Glen Gery) Low Btu Gasification*. U.S. Environmental Protection Agency Report EPA–600/7–79–185. Washington, DC.

Townsend, W. N. and Gilham, E. W. F. (1975). Pulverized fuel ash as a medium for plant growth. *The Ecology of Resource Degradation and Renewal* (ed. by M. J. Chadwick and G. T. Goodman), pp. 287–304. The 15th Symposium of the British Ecological Society, Blackwell, Oxford.

Turekin, K. K. and Wedepohl, K. H. (1961). Distribution of the elements in some major units of the earth's crust. *Bulletin of the American Geological Society*, **72**, 175–91.

UNGA (1979). United Nations General Assembly. Technologically Modified Exposures to Natural Radiation. United Nations Publication No. A/AC.82/R.358.

UNGA (1980). United Nations General Assembly. Technologically Modified Exposures to Natural Radiation. *United Nations Scientific Committee on the Effects of Atomic Radiation.* Vienna. UN Publication No. A/AC.82/R.371.

UNSCEAR (1977). *United Nations Scientific Committee on the Effects of Atomic Radiation*. 1977 Report on the General Assembly. United Nations publication No. E.77. ix.1, New York.

U.S. National Committee for Geochemistry (1980). *Trace-Element Geochemistry of Coal Resource Development Related to Environmental Quality and*

Health. Assembly of Mathematical and Physical Sciences, National Research Council. National Academy Press, Washington, DC.

VAN HOOK, R. I. (1978). Potential health and environmental effects of trace elements and radionuclides from increased coal utilization. Oak Ridge National Laboratory, Tennessee. Report No. ORNL–5367.

VINOGRADOV, A. P. (1959). *The Geochemistry of Rare and Dispersed Chemical Elements in Soils*, 2nd edn. Consultants Bureau, Inc.

WADE, N. (1979). Synfuels in haste, repent at leisure. *Science*, **205**, 167–8.

WALSH, P. R., DUCE, R. A. and FASCHING, J. L. (1979). Considerations of the enrichment, sources and flux of arsenic in the troposphere. *Journal of Geophysical Research*, **84**, 1719–26.

WANGEN, L. E. and WILLIAMS, M. D. (1978). Elemental deposition downwind of a coal-fired power plant. *Water, Air & Soil Pollution*, **10**, 33–44.

WIERSMA, G. B. and CROCKETT, A. B. (1978). Trace elements in soil around the Four Corners power plant. Environmental Protection Agency Report, 600/3–78–079.

YAVORSKY, P. M. and AKHTAR, S. (1974). *Proceedings of the Symposium on Environmental Aspects of Fuel Conversion Technology*, St Louis, EPA–650/2–74–118, 325–30.

ZOLLER, W. H., GLADNEY, E. S., GORDON, G. E. and BORS, J. J. (1974). Emissions of trace elements from coal-fired power plants. *Trace Substances in Environmental Health* VIII (ed. by D. D. Hemphill); pp. 167–72. University of Missouri, Columbia, Missouri.

ZUBOVIC, P., SHEFFEY, N. B. and STADNICHENKO, T. (1967). Distribution of minor elements in some coals of the Western and South Western regions of the Interior Coal Province. *U.S. Geological Survey Bulletin*, 117–D.

7

The Environmental Significance of Coal-derived Carbon Compounds

1 Introduction

Many carbon compounds, including coal derivatives, formed in manufacturing processes, possess high biological and pathological activities. Polycyclic aromatic hydrocarbons (PAH) are of considerable interest in this respect as many are known to be mutagenic and some have been shown to be carcinogenic in animal systems.

Polycyclic aromatic hydrocarbons can originate from natural sources or be emitted in the course of a wide range of anthropogenic activites. These include aluminium plants, coke plants, iron and steel works, ferroalloy plants and foundries, automobiles and other petroleum-burning engines (including aircraft) and waste disposal by burning (Jones, 1979; Ahland and Mertens, 1979). However, relatively long-term monitoring of PAH concentrations in the Netherlands (Brasser, 1979) has revealed that during the period that natural gas replaced coal and oil as a fuel for heating buildings (and to some extent in factories and power stations) there was a spectacular fall in concentrations from relatively high to negligible levels. This fact, and the data summarized by Suess (1976) and given in Table 7.1, therefore make it natural that PAH should be one focus of attention for the possible health impacts of coal combustion and coal conversion technologies.

In view of the interacting environmental factors, and range of sources of PAH emissions, the impact of coal conversion technologies is assessed against a historical background and in the light of a wide range of other anthropogenic activities.

2 Human Health Effects of Coal-derived Carbon Compounds

2.1 Early studies on coal tar

The prominent English surgeon P. Pott described, in 1775, an occupational disease of chimney sweeps. He correctly ascribed skin cancer affecting the scrotum and thighs to the continual exposure of the tissue to soot. These observations showed that there was a close correlation between coal products and malignant growths in man. This has subsequently been confirmed as a result of studies of occupational cancer in man in other industries: for example, from contact with unrefined paraffins; in workers making briquets from coal dust (Table 7.2); in workers in continual contact with petroleum products and in other situations. Skin tumour formation appears to

TABLE 7.1 *Estimated benzo(α)pyrene emissions to the atmosphere (Suess, 1976)*

Sources of formation	BP emissions in t yr^{-1} U.S.A.	World-wide excluding U.S.A.	World-wide
Heating and power generation			
Coal	431	1945	2376
Oil	2	3	5
Gas	2	1	3
Wood	40	180	220
TOTAL	475	2129	2604
Industrial processes			
Coke production	192	841	1033
Catalytic cracking	6	6	12
TOTAL	198	847	1045
Refuse and open burning			
Enclosed incineration			
Commerce and industry	23	46	69
Other	11	22	33
Open burning			
Coal refuse	340	340	680
Forest and agriculture	140	280	420
Other	74	74	148
TOTAL	588	762	1350
Vehicles			
Trucks and buses	12	17	29
Automobiles	10	6	16
TOTAL	22	23	45
GRAND TOTAL	1283	3761	5044

be directly dependent on the duration of contact with the carcinogenic agent which, in the case of the coal-briquet makers, was pitch and tar. Since the technology of briquet-making has been changed in order to eliminate skin contact, skin tumours have disappeared.

2.1.1 Experimental work with coal tar

Systemic application of coal tar to the skin of the ears of rabbits for a year was found to induce, initially, papillomas and then carcinomas that metastasized into the lymph glands (Yamagiwa and Ichikawa,

TABLE 7.2 *Development of professional skin tumours in the briquet production industry (Teutschlaender, 1929)*

No. of years of work	No. of patients	Percentage of patients with skin tumours (papillomas, cancer)
1–5	40	17
5–10	38	21
10–15	22	36
15–20	9	67
20–25	13	69
25–30	14	71
30–35	13	77
35–40	5	80
>40	4	100

1918); this has been confirmed by other workers using different animal species (Petrov *et al.*, 1960). It was then established that coal tars differ in their carcinogenicity; for instance 'Limburg coal tar' was found not to be carcinogenic at all. Coal type and the temperature of coking were found to be of importance in this respect. Coal tar fractions obtained at different temperatures had different carcinogenic properties.

Coal tar is a complex mixture of a larger number of aromatic hydrocarbons (benzene, naphthalene, anthracene, phenanthrene and their homologues), aromatic acids and nitrogen-containing compounds and it transpired that it was better to study the carcinogenic effect of separate fractions obtained by distillation at different temperatures. It became established that only the tar fractions distilled at 400–600°C and above are carcinogenic; distillation at 800–850°C results in the highest yield of carcinogenic products. It was concluded that carcinogenic products have a high boiling point, are stable when heated repeatedly, do not contain sulphur or arsenic, are neither an acid nor an alkali and are not nitrogen-containing compounds. These characteristics placed the neutral, polynuclear aromatic hydrocarbons with a high boiling point under suspicion as the carcinogenic agents. A study of the actions of sulphuric and picric acids on the different coal tars and schist resins suggested that the carcinogenic principle in tars is found amongst the higher polynuclear hydrocarbons; the treatment of certain tars with sulphuric acid was found to deprive them of carcinogenic activity.

2.1.2 The discovery of chemically pure carcinogens

Carcinogenic tars could be obtained from various materials by heating them to 560–880°C; for example, tars capable of producing skin cancer in mice were obtained by heating isoprene to 820°C or acetylene to 700°C. These substances had to be distilled in a hydrogen atmosphere and it was established that the tars obtained represented a mixture of higher hydrocarbons.

In 1930 the fluorescence of a number of known carcinogenic tars and mineral oils was described (in ultraviolet light, a spectrum with three bands at 400, 418 and 440 nm was seen). Since benz(α)-anthracene and a number of other substances from the same group possess a similar fluorescence, this led to a range of polycyclic aromatic hydrocarbons (PAH) being investigated for carcinogenic properties, among them dibenz(α,h)anthracene (Cook *et al.*, 1932). Painting the skin of mice with a 0.3% solution of this substance in benzene resulted in almost half of the animals developing, after 6 months, firstly papillomas and then carcinomas. Injection of several mg of this substance under the skin of mice and rats led to the development of sarcomas in the majority of animals treated.

Comparison of the carcinogenic activity of dibenz(α,h)anthracene with that of certain well-known types of coal tar showed that the latter possessed a more pronounced action and hence contained a more powerful carcinogenic substance. After persistent and pains-taking work, Cook, Hewitt and Hieger (1933) succeeded in isolating this substance from coal tar and identifying it as benzo(α)pyrene. This hydrocarbon induced cancer in mice in half the time taken by dibenz(α,h)anthracene and in a wider range of animal species. Benzo(α)pyrene (BP) was subsequently prepared synthetically.

2.2 *Carcinogenic Polycyclic Aromatic Hydrocarbons (PAH)*

The names and structural formulae of several of these compounds are shown in Figure 7.1.

DBA was synthesized in 1929 and first tested in 1930–32 (Cook *et al.*, 1932). It possesses average carcinogenicity, producing tumours in many species of laboratory animals, even in animals refractory to PAH, such as the guinea pig.

DMBA was first synthesized in 1938 independently in the U.S.S.R. and in the U.S.A., and when tested it proved to be the most important of all the carcinogenic PAHs. Painting the skin of mice with a 0.05% solution of DMBA in benzene leads to the appearance of papillomas on or around the 40th day of treatment and, in experiments using a 0.1% solution, papillomas appeared within 2–3 weeks. Painting the skin of the ears of rabbits with a 1% benzene solution of DMBA produced cancer in almost all the animals in 4 months.

BP is probably the best known and most intensively investigated carcinogenic PAH; it has been tested in a whole range of laboratory animals and its carcinogenic potency is about twice that of DBA.

DBP, which was synthesized in 1937, shows a carcinogenicity approximately equal to that of BP when tested in mice.

MC was synthesized in the U.K. and the U.S.A. from desoxycholic acid, a normal component of human and animal bile. MC is approximately equal to BP in its carcinogenic properties but when applied to the skin of mice, MC elicits a more prominent inflammatory reaction.

The carcinogenicity of cholanthrene is slightly lower than that of MC.

2.2.1 Carcinogenic activity and PAH structure

Not all polycyclic aromatic hydrocarbons are carcinogenic, and those that are differ widely in terms of potency. A great deal of effort has therefore been expended in trying to discover the scientific basis for the structure-activity relationships that exist within the hydrocarbons as a class of chemical carcinogens. In addition, the mechanism of action of PAHs as carcinogens has attracted considerable attention, not least because it was difficult to envisage quite how these compounds, which are commonly formed at high temperatures and are relatively inert chemicals, might change normal into malignant cells. So far, somewhat better progress has been made in the second of these areas of interest than has occurred in the first.

The K-regions of PAHs (Figure 7.1), which were known to be the most chemically reactive and electron-dense regions of hydrocarbon molecules, initially attracted much attention (Pullman and Pullman, 1955) and this was followed by a period in which the metabolism of the hydrocarbons was intensively investigated (for a review see Sims and Grover, 1974). It was known that the aromatic double bonds present in the molecules could be metabolized (Figure 7.2) by the so-called drug-metabolizing enzymes that are widely distributed throughout the animal kingdom and that enable many species to transform foreign compounds into water-soluble derivatives that can be excreted. Moreover it was also found that, when PAHs were metabolized by these enzymes, *in vitro*, reactive intermediates were

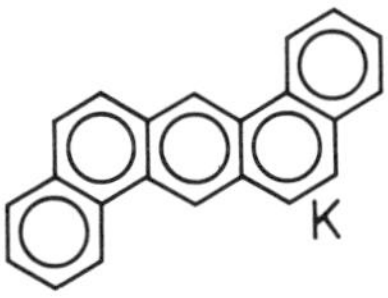

DBA – dibenz(α,h)anthracene
(1,2,5,6,–dibenzanthracene)

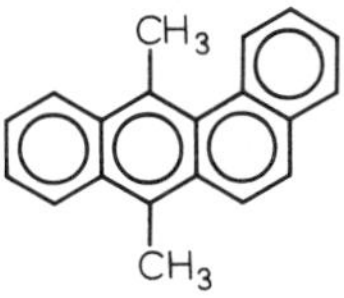

DMBA – 7,12–dimethylbenz(α)anthracene
(9,10–dimethyl–1,2,–benzanthracene)

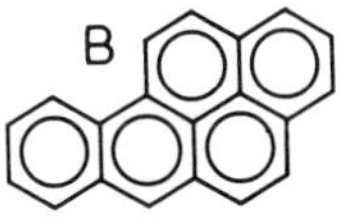

BP – benzo(α)pyrene
(3,4,–benzpyrene)

DPB – dibenzo(α,i)pyrene
(3,4,8,9–dibenzpyrene)

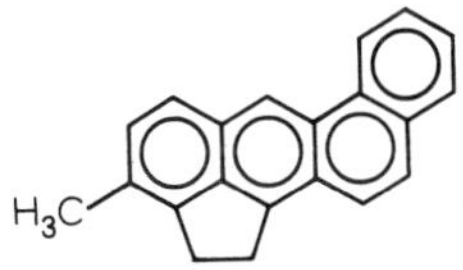

MC – 3–methylcholanthrene
(20–methylcholanthrene)

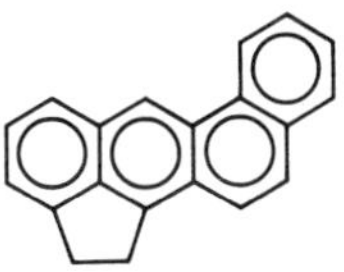

C – cholanthrene

FIGURE 7.1 *The structural formulae of some important carcinogenic polycyclic aromatic hydrocarbons (the older nomenclature is given in parentheses) showing the "K-region" (K) and the "bay-region" (B)*

formed that could react covalently with, for example, DNA and protein. Initially these intermediates, which of course might also be formed in susceptible tissues *in vivo*, were thought to be simple epoxides. However, over the last few years it has become apparent that the reactive species are most probably diol-expoxides (Figure 7.2). The particular diol-epoxides that are most reactive are diol-epoxides of the "bay-region" type in which the epoxide moiety is adjacent to a "bay" in the molecule (Figure 7.3), i.e. the region formed by the presence of an angular benzene ring. Metabolites of this type are currently under active investigation (see review by Grover, 1980) since differences in their properties and in their rates of formation and further metabolism may help to explain the very marked structure-activity relationships that exist within the PAHs.

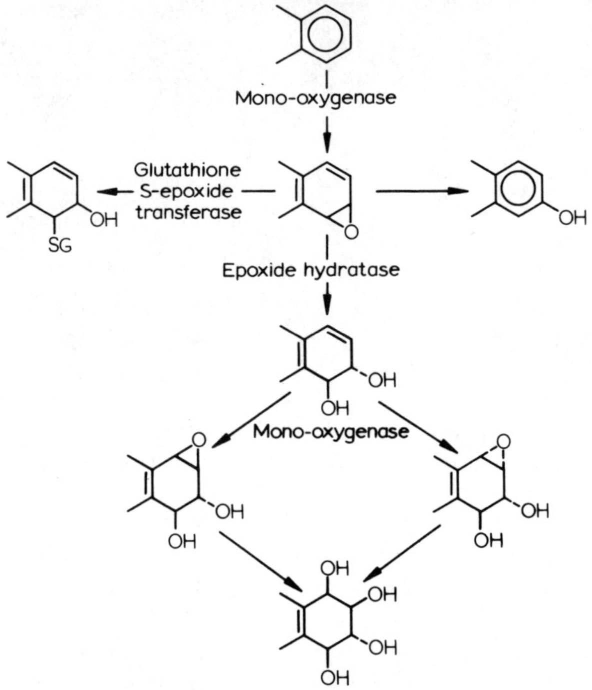

FIGURE 7.2 *The metabolism of an aromatic ring in a polycyclic hydrocarbon to hydroxylated derivatives*

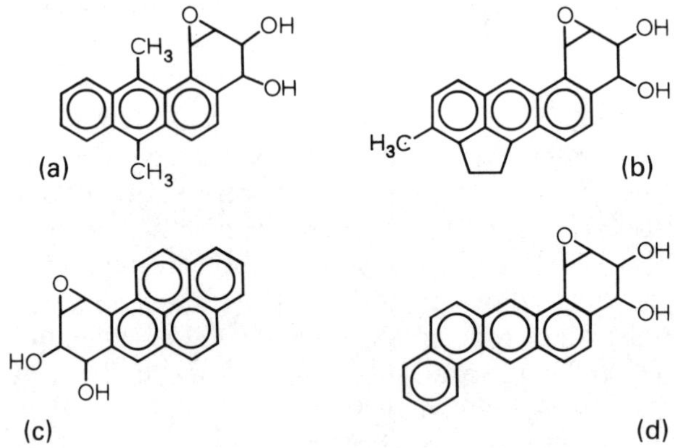

FIGURE 7.3 *Reactive "bay-region" diol-epoxides of (a) 7,12–dimethylbenz(α)anthracene, (b) 3–methylcholanthrene, (c) benzo-(α)pyrene, (d) dibenz(α,h)anthracene*

2.2.2 Benzo(α)pyrene (BP)

Even when present in complex mixtures, BP can be determined quantitatively by relatively precise and easily performed methods (Shabad *et al.*, 1972; Egan *et al.*, 1979). Spectro-luminescence methods, in which fine-structured spectra are obtained at low temperatures (−196°C) (Khesina, 1979; Yanisheva and Kireeva, 1979), and methods based on gas chromatography (Grimmer and Bohnke, 1979) have frequently been employed.

BP is the most widely distributed PAH carcinogen found in the environment. Of the several thousand samples of air, soil and water that have been examined in the U.S.S.R., there was none from which it was absent. Samples rich in BP usually also contain relatively large quantities in the exhaust gases of automobiles, aeroplanes and motorized ships. Benzene extracts of condensates of these gases, and of soots, produce papillomas and skin cancers in mice with the same concentration (Shabad, 1973). Painting the skin of mice with the products of the distillation of coal tar and of petroleum containing different concentrations of BP gives neoplasms that correspond in number to the BP concentration used (Yanisheva, Chernichenko and Balenko, 1977).

In a study of automobile emissions (Huenigen *et al.*, 1978), a relationship was established between the concentration of BP and the concentration of other PAH compounds that were present. It is therefore possible to calculate the approximate quantities of the other PAH compounds present from BP estimations, although for some types of automobile exhausts the relationship does not hold.

Of late, many new forms of occupational tumours have been described. These are often due to exposure of products containing large quantities of BP that are formed in the partial combustion of fossil fuels. Lawther & Waller (1976) found that roofing workers who had sustained contact with material containing coal tar and pitch, contracted skin cancer 4 times and lung cancer 1.5 times more frequently than control groups. Fox, Lindars and Owen (1974) have reported that a higher incidence of malignant disease of the respiratory organs is seen in workers in the rubber and cable industries, and Konstantinov *et al.* (1974) have described skin cancers in workers in aluminium electrolysis plants. In all these cases the contact with PAH in general, and with BP in particular, was over a long period. BP was not the only carcinogenic agent present and it probably acted in combination with other PAHs. It seems reasonable, however, to use BP as typical of the entire group.

Testing both coal tars isolated from industrial products and wastes

TABLE 7.3 *Development of tumours in mice ($CC_{57}W$ strain) painted with treated products of coal tar and petroleum with different BP contents (Yanisheva, Chernichenko and Balenko, 1977; Yanisheva and Kireeva, 1977)*

Material studied	Mean BP content (%)	No. animals surviving till the first tumour	No. of animals with tumours: Skin			Other organs			Systemic lesions (leukaemia)
			Total	Malignant	With metastases	Total	Malignant	With metastases	
Coal tar pitch	0.24	43	38	35	9	24	1	1	0
Petro-bituminous compounds from cracking residue									
BH–5	0.0138	49	9	6	3	11	9	0	2
BH–4	0.0027	42	4	2	0	4	0	0	0
Petro-bituminous compounds from petroleum tars									
BH–5	0.00002	43	2	1	0	5	3	0	0
BH–2	0	40	0	0	0	1	1	0	0

produced by various industries, shows that carcinogenic activity is correlated with the physicochemical properties of BP. In addition, the greater the BP concentration present in the product studied, the higher seem to be the number of neoplasms that occur in treated animals, although it is too early yet to assume a simple linear dose-response curve or to assume zero repair. The tumours induced include malignant tumours capable of metastasizing (Tables 7.3 and 7.4).

BP can remain unchanged in the soil for long periods and laboratory and field experiments have shown that the time required for complete destruction (99%) of BP in the soil is from 3 to 5 years. The exact fate of BP released to the atmosphere is unclear. It was, for some time, assumed that it was extremely stable, but photochemical decomposition is now an accepted fate (Grimmer, 1979). However, recently, stabilization of benzo(α)pyrene when adsorbed on to coal fly ash has been investigated (Korfmacher *et al.*, 1980). A number of adsorbed polycyclic aromatic hydrocarbons will spontaneously oxidize in the absence of light. The decomposition rate appears to be fly ash-dependent.

A comparative study of different coal and wood tars that have been employed in the preparation of tar ointments (Shabad *et al.*, 1970) showed that their carcinogenicity in mice corresponds to their BP content, the strongest carcinogenic activity being seen after painting the skin of mice with lianthrol (CIBA, Switzerland), used in the preparation of "Locacorten-Tar". However, this straightforward relationship has recently been questioned (Henschler, 1980).

TABLE 7.4 *Carcinogenic properties of waste products of the coal processing industry (Yanisheva, 1939)*

Coal tar under study	Months to appearance of the first tumour	Percentage of animals with tumours		BP concentration in coal-tar (%)
		Malignant	Benign	
From coke gas	2	95.4	4.6	0.1
Exudate from coke furnaces	4	86.6	13.4	0.06
From atmospheric dust (near coking factory)	6	44.4	38.9	0.03
From phenol sediment collectors	2	52.7	19.1	0.03

Investigations (Klipinitser *et al.*, 1977) showed that the anthracene fraction of coal tar contains especially high quantities of PAH and BP in particular (up to 27 000 $\mu g\ kg^{-1}$). However, BP content is markedly decreased in SANT-H tar and other products obtained by continuous combustion. This is an example of the advantages that may be achieved by improved technology which are essential since polycyclic organic matter (POM), derived from the combustion of fossil fuels, mainly coal, may cause 5–10 cases of lung cancer per 100 000 males yr^{-1}. An enhancement of 1 μg 1000 m^{-3} in atmospheric POM gives a 50% increase in deaths from lung cancer. In some situations, BP concentration can provide a good indication of POM concentrations (Shabad, 1978).

2.2.3 Environmental sources of PAH

The main sources of carcinogenic hydrocarbons are heating systems, emissions from industrial and energy-producing plants, automobile emissions and aviation exhausts and these are considered separately below.

Heating systems

In 1948 a study was begun of atmospheric PAH in three cities in the U.S.S.R.: Moscow, Ivanovo and Grozny. These cities were chosen because coal was the main fuel used for heating in Moscow, wood predominated in Ivanovo, whilst liquid fuels were the main heat source in the third city. It was found that the quantity of BP in the atmosphere depends not so much on the type of fuel as on the methods of combustion used and improved combustion decreases the quantity of BP emitted into the atmosphere. These studies established, therefore, a basis for prophylactic measures.

Improvements in combustion and in smoke control can be gained by replacement of individual heating systems with regional and central heating installations combined with electricity generation. The use of gas in place of other types of fuel, increased electricity generation and neighbourhood heating schemes all represent real measures that can be taken against atmospheric pollution. The centralized heat supply of cities by large thermo-electric stations has become a decisive measure in enabling the closure of thousands of small boilers with unfavourable conditions for coal burning and the substitution for them of a few thermo-electric stations with favourable coal-burning conditions and better emission control. Rational city planning, which separates industrial zones from residential areas, and the creation of "green zones" also bring improvements and much has been achieved in Moscow, and in other large cities, through improvements such as these.

Industry

For certain cities and regions, it is industrial emissions and automobile exhausts rather than domestic heating systems that make the greatest contribution to atmospheric pollution. A pitch coking plant in the village of Buroz, for example, was responsible for much of the atmospheric pollution. Changes in plant design, giving improved combustion characteristics, decreased the quantity of smoke, pitch and BP emitted from this plant (Shabad and Dikun, 1959).

Investigations of a number of metallurgic, oil refining, pitch coking and briquet manufacturing plants confirmed the dependence of the quantity of carcinogenic hydrocarbons emitted, on the combustion characteristics (Yanisheva, Chernichenko and Balenko, 1977). The main measures to be taken to reduce atmospheric pollution by carcinogenic substances, must therefore be technological improvements in plant design and operation. A considerable decrease in atmospheric pollution in London has been brought about by improvements in the design of heating systems and industrial plants, and by restrictions that have been placed on the types of fuel that may be used (Lawther and Waller, 1976). The quantity of BP present in the atmosphere in London has decreased tenfold in a 25-year period.

Automobile emissions

The study of automobile exhaust gases is of extreme importance to the assessment of coal conversion effects, as these can be a main source of atmospheric pollution with carcinogenic substances. Waller (1952) showed that automobile exhaust gases contain BP and in the U.S.S.R., carcinogenic substances have been detected in soot and in exhaust gases from automobiles, by many workers (Gurinov, Tugarinova and Vasilieva, 1962; Gofmekler *et al.*, 1963; Shabad, Khesina and Khitrovo, 1968; Zabezhensky, 1968; Varshavsky *et al.*, 1970). Atmospheric BP concentrations on highways exceed the background concentration by a factor of at least 1.5 and it can be shown that, with a slight lag period, BP concentration is positively correlated with traffic density.

The mean concentration of BP in the air on main roads is approximately 3 μg 100 m^{-3}. If the volume of air inhaled by a man in 24 hours is 20 m^3, then the inhabitants of a large city can inhale about 0.6 μg BP day^{-1} from automobile exhaust gases. Roughly the same quantity of BP is contained in the smoke of 50 cigarettes. Studies in a small town in Austria showed that from 1958 to 1970, mortality from cancer among people living near the highway was 9 times higher than that occurring amongst people living some way away from it, in these particular studies. Soil near the highway also had a PAH content that was much higher than that present in soil distant from it. Improvements in engine design are required in order to reduce PAH emissions in exhaust gases (Huenigen *et al.*, 1978; Shabad, Khesina and Khitrovo, 1968).

Aviation exhausts

The presence of carcinogenic hydrocarbons in the exhaust gases of automobile engines suggested that other internal combustion engines using the same fuel should be investigated. It has been shown that the soot from exhaust pipes of aviation engines and the exhaust gases themselves contain BP (Shabad and Smirnov, 1976). The gas turbine engines of modern aircraft emit exhaust gases with 2–4 mg BP min^{-1} and many times more is present during take off (up to 40 mg).

Extracts from soots formed by aviation engines induce malignant tumours in mice in 100% of cases, and soots from piston engines induce neoplasms 2 weeks earlier than do soots from turbo-jets. All the tumours formed initially were papillomas and these became transformed later into epidermoid cancers; in some cases carcino-sarcomas and sarcomas were seen.

BP has also been detected in the plants and soil around Moscow airport, the quantities reducing with increasing distance away from the centre of the airport. Dust collected from the runway itself contained 182 μg kg^{-1} of BP.

3 Polycyclic Aromatic Hydrocarbons in the Food Chain

3.1 Atmospheric PAH

The sources of atmospheric PAH have already been outlined. Aspiration samples taken in cities in the U.S.S.R. show that concentrations vary considerably. Generally, the larger the city, the higher the concentration and levels of PAH, and concentrations in industrial regions exceed those present in rural areas. Especially high concentrations of PAH are found associated with coal and petroleum utilization.

In the U.S.S.R., PAH levels are generally higher in winter than summer, with the exception of certain large cities, where seasonal variations in traffic densities (higher in summer) override this trend.

Certain factors can help in removing PAH from the atmosphere. Ultraviolet radiation and atmospheric ozone seem to cause the destruction of carcinogenic PAH, and particulate and aerosol sedimentation also result in their removal (Shabad and Dikun, 1959; Shabad, 1973).

3.2 *PAH in soils*

Rainout, washout and sedimentation leads to the addition of carcinogenic hydrocarbons to the soil, particularly in the vicinity of large industrial towns. Shabad and Dikun (1959) demonstrated the presence of benzo(α)pyrene in a park zone in Leningrad. Blumer (1961), Borneff and Fabian (1963), Mallett and Héros (1962), and Zdrazil and Picha (1966) have similarly demonstrated the presence of BP in soils in the U.S.A., the F.R.G., France and Czechoslovakia respectively. Yanisheva and Kireeva (1977) were able to demonstrate that there was a relationship between soil BP concentration and distance from industrial sources, particularly coke-making plants.

Losses of BP from the soil take place mainly during spring and summer following increased deposition during the winter. Leaching occurs from the upper down to the lower layers of the soil (Table 7.5).

Plant uptake, through the roots, can also account for some reduction in soil concentration and, in addition to decomposition by ultraviolet radiation, bacterial breakdown also occurs.

3.2.1 Contamination by soil treatment

The use of irrigation water from contaminated sources can add BP to soils but by far the most noticeable effects result from the treatment of soil with certain hydrocarbon products in order to prevent erosion. In the U.S.S.R., "Nerosine", an Estonian oil shale product, is added to soil as an anti-erosion treatment and the BP content of this product may vary from 30 to 100 ppm. It has been

TABLE 7.5 *Distribution of benzo(α)pyrene in soil layers, in μg kg^{-1} (Shabad, 1973)*

Depth of layer (cm)	Distance from oil-refining plant 250 m	500 m	1500 m
2.5	11.90	1.20	0.12
5.0	12.97	1.24	0.11
10.0	11.08	1.95	0.13
20.0	17.34	1.62	0.09
30.0	14.53	1.12	0.008
50.0	13.53	0.081	0.001
80.0	2.77	0.18	0.002
110.0	5.40	0.002	0.0005
150.0	–	0.002	–

TABLE 7.6 *BP content (μg kg^{-1}) in soil samples treated with Nerosine in May 1969, in (a) Cotton–Uzbek S.S.R. and (b) Spring wheat–Kazack S.S.R. (Shabad, 1973)*

(a)

Soil depth (cm)	Control		Nerosine 1.4t ha^{-1}		Nerosine 2.1t ha^{-1}	
	May	Sept	May	Sept	May	Sept
0–10	4.7	8.2	64.0	47.0	235.0	153.0
10–20	4.9	2.8	15.0	47.0	46.0	46.0
20–30	4.4	5.0	32.0	54.0	9.0	79.0

(b)

Soil depth (cm)	Control				Nerosine-treated			
	1969			1970	1969			1970
	June	July	Aug.	Aug.	June	July	Aug.	Aug.
0–10	0.8	1.0	3.5	1.0	238.0	186.0	169.0	37.0
10–20	0.3	1.0	1.2	0.8	95.0	42.0	14.0	7.2
20–30	0.1	0.3	1.2	0.1	9.0	25.0	21.0	5.2

shown that Nerosine is carcinogenic in mice. After application to the soil surface, BP from the product rapidly penetrates deeper soil layers (Table 7.6).

3.3 PAH uptake by plants

Plants grown on soils with high PAH concentrations absorb BP through the roots and translocate it to the shoots. Asters (*Aster nova-belgii*) and nasturtiums (*Tropaeolum majus*) grown on soil highly polluted with BP contained 6.3 times as much BP as the same species grown on control soils. However, no significant elevation of BP levels was found in the grain of cereals (Winter and Spring wheat, rye, barley or maize) grown on Nerosine-treated soil, but treatment does result in BP accumulation in sunflower (*Helianthus annuus*) seeds.

3.3.1 PAH levels in food

There is evidence of relatively high concentrations of PAH (particularly BP) in some food products (Fritz *et al.*, 1974). Potatoes may contain up to 1 $\mu g\ kg^{-1}$ and even higher amounts are present in samples from some particular localities (Gräf and Diehl, 1966; Grimmer and Hilderbrandt, 1967), usually in the skins. In other food products from Georgia (particularly vegetables and fruit), BP concentrations were seldom above background levels (Ilnitsky, Belitsky and Shabad, 1975). Levels in soil in excess of 100 $\mu g\ kg^{-1}$ may give rise to significant increases in BP concentrations in vegetables.

Table 7.7 gives details of BP levels in soils and vegetable products in relation to locality. The effects of proximity to a highway are evident. The concentrations present in vegetables depend not only on location but also on deposition characteristics, soil factors and the crop species concerned. The highest concentrations were found in cabbages (*Brassica oleracea*) and the lowest in pumpkins (*Cucurbita pepo*). Particulate deposition on the leaves of cabbages accounted for a high proportion of the concentration.

The mean cumulative dose of PAH to a 70-year-old human might be 3 mg from food (excluding vegetables), 2 mg from vegetables and 0.3–0.4 mg from drinking-water. In addition, some PAH might accrue from food packaging, processing and preparation.

TABLE 7.7 *BP content (μg kg^{-1}) in soils and vegetables grown in different areas of the Ukraine (Yanisheva and Kireeva, 1977)*

Population point	In the soil		In vegetables		Dose of benzo(α)-pyrene entering the human body in vegetables over 70 years (mg)
	Near the highway	Among houses	Near the highway	Among houses	
Large administrative centre	1937±306	249±70	0.67±0.09	0.17±0.04	3.1−12.1
Average industrial town	348	15±5.5	0.21±0.08	0.11±0.04	2.0±3.8
Small town	160.6	38±11.4	–	0.16±0.02	2.9
Settled agricultural area	32.0	18.7±4.5	–	–	–
Uninhabited area	5.0±1.6	–	–	0.10±0.02	1.8

3.4 PAH in water

3.4.1 Freshwater reservoirs

Systematic studies of the fate of BP and other PAH compounds in freshwater reservoirs have not been undertaken, although some information is available (Ilnitsky, Voronin and Solenova, 1974; Ilnitsky, Belitsky and Shabad, 1975; Shabad, 1973). The main sources of PAH in reservoirs are: exhaust gases from the engines of boats, effluents from industrial plants, urban run-offs from cities and seepage of sub-soil water.

Comparison of BP content in plankton, bottom sediments and higher water plants in two water reservoirs near Moscow showed that the reservoir where the passage of boats was authorized had 2–10 times more BP than the one where such activities were forbidden.

Table 7.8 shows the significance of sewage water for pollution of a large river with BP.

Table 7.9 shows the effect of the input from a large city into a river. Concentrations in the water, in aquatic plants and in bottom sediments are many times higher than they are in the river upstream of the city, the contamination presumably resulting from industrial emissions and urban run-off.

As an indication of PAH contamination in reservoirs, BP concentrations were used. BP possessed a very low solubility in water (about 0.1 $\mu g\ l^{-1}$) but organic solvents (acetone, petroleum and petroleum products, detergents) increase this solubility; the stability

TABLE 7.8 *BP concentrations in the water of a large river in 1971–74 (Lembik, 1979)*

Position	No. of samples	BP content ($\mu g\ l^{-1}$)
26 km upstream from the inflow of sewage water	32	0.0130±0.0002
2 km upstream from the inflow of sewage water	36	0.0110±0.0008
500 m downstream from the inflow of sewage water	36	0.0250±0.0008
10 km downstream from the inflow of sewage water	36	0.0120±0.0005

TABLE 7.9 *BP concentration in a river upstream from and downstream from a large city (Shabad, 1973)*

	Upstream	Downstream
Water ($\mu g\ l^{-1}$)	0.001	0.04
Plankton ($\mu g\ kg^{-1}$)	–	475
Aquatic plants ($\mu g\ kg^{-1}$)	0.0–6.0	75
Bottom sediment ($\mu g\ kg^{-1}$)	18	146–403

of BP in water is sufficient to enable it to spread considerable distances. After 30–45 days' storage, 5–20% of the initial concentration of 10 $\mu g\ l^{-1}$ remained in both tap- and lake-water.

The BP concentration in reservoirs in industrial regions is higher than in rural districts, and a study of several dozen bodies of water (rivers, lakes and reservoirs) in the U.S.S.R. permitted a number of significant trends to be recognized (Table 7.10).

The pollution of freshwater reservoirs with PAH raises the question of preventing such substances from reaching drinking-water supplies. It is important to remove tars and other products that contain PAH from the effluents coming from industrial plants and cities. Further, it is necessary to employ both biological methods of purification and filtration through activated charcoal. After these initial stages of purification, water must also be subjected to the action of chlorine or ozone.

Ilnitsky *et al.* (1968) concluded that chlorination of water is relatively ineffective as far as PAHs (especially BP) are concerned. Comparison with the actions of ultraviolet light and ozone showed that the latter destroys BP and a number of other PAHs many times more effectively and more completely than ultra-violet light which,

TABLE 7.10 *Degree of pollution by BP in water bodies (Shabad, 1973)*

	Background level	Degree of pollution with BP		
		Moderate	Considerate	High
Water ($\mu g\ l^{-1}$)	Up to 0.0003–0.0007	Up to 0.005	Up to 0.01	Above 0.01
Bottom sediments ($\mu g\ kg^{-1}$)	Up to 1.0–3.0	Up to 20–30	Up to 100	Above 100
Higher water plants ($\mu g\ kg^{-1}$)	Up to 1.0–3.0	Up to 15–20	Up to 50	Above 50

in turn, is more effective than chlorination. Burleson, Caulfield and Pollard (1979) found that ozonization removes mutagenic and carcinogenic effects that are due to the presence of PAH and cyclic amines in water, as well as inactivating certain mutagenic pesticides.

3.4.2 Seas and oceans

The growing use of seas and oceans as a source of potable water (after distillation) emphasizes the need to consider their pollution with PAH. The pollution of sea and ocean waters emanates both from general sources (shipping, aviation, atmospheric particulates) and from specific or local sources (specific catastrophes or outflows from industrial plants engaged, for example, in coal conversion or in shale and oil refining). With regard to environmental catastrophes, a modern oil tanker in a limited sea area can deliver instantaneously several kilograms of BP and other PAHs. The polluting effect of a shore industry can be clearly seen from estimations of BP in sand, aquatic plants and other littoral organisms, and Gortalum (1972) found higher levels of pollution by BP in the Baltic than in the Black Sea, in the region of Yevpatoriya. Andelman and Suess (1970) reported on BP concentrations in marine plants and also found considerable quantities of BP in fish, molluscs, oysters and mussels. According to Gortalum (1972), BP accumulates in especially large quantities in jellyfish (over 71.7 $\mu g\ kg^{-1}$ dry weight) and this is thought to be due to PAH absorption from the sea water.

Ermakov (1979) took 300 samples from different seas in and around the U.S.S.R. Special attention was paid not only to the bottom layer and to water taken at some depth, but also to the so-called border layer. This is a film forming a border between two phases (atmosphere and hydrosphere) where different pollutants accumulate. BP was found in varying quantities with considerable differences in mean pollution levels for different seas becoming evident (Table 7.11). The Bering Sea concentrations can be considered as background levels, and Baku Bay and the estuary of the River Purtse were found to be the two most polluted. This is undoubtedly due to the presence of the oil industry and its refineries and to the shale oil industry, respectively. The Bay of Finland is considerably more polluted than the central part of the Baltic Sea, as it is adjacent to industrial areas.

On the basis of the study it is possible to assess, semi-quantitatively, the different levels of marine pollution using BP as an indicator of total PAH. Such an assessment is given in Table 7.12.

TABLE 7.11 *Mean concentration ($\mu g\ l^{-1}$) of BP in seas in and around the U.S.S.R. (Ermakov, 1979)*

Object of study	Water	Surface layer	Bottom sediment
Bering Sea (central area)	0.025±0.003	0.125±0.003	–
Baltic Sea (central area)	0.051±0.004	–	16.1±1.17
Baltic Sea (Bay of Finland)	0.15±0.01	2.24±0.1	76.8±3.25
Baltic Sea (estuary of the River Purtse)	0.27±0.07	5.0±0.25	7500±125
Caspian Sea (Bay of Baku)	0.10±0.02	10.6±0.39	8030±146

TABLE 7.12 *Pollution levels ($\mu g\ l^{-1}$) in the marine environment by BP (Ermakov, 1979)*

Degree of pollution	Concentration of BP in water	Concentration of BP in the surface layer	Concentration of BP in the bottom sediment
High	>0.1	>0.1	>100
Average	0.01–0.1	0.1–1.0	20–100
"Background"	<0.01	<0.1	<20

3.5 *Bacterial degradation of PAH*

The decomposition of polycyclic aromatic hydrocarbons was first demonstrated many years ago by Tauson and Veselov (1934) and this has been confirmed more recently in work on the degradation of crude oil in sea water by micro-organisms (Zo Bell, 1969). Isolates of soil bacteria (*Bacterium megaterium* and *B. sphericus*) have proved capable of destroying BP in soil by metabolizing it, and it has also been shown that this leads to a reduction in carcinogenic activity in mice. In soil, the seasonal variations in soil BP content can be related to seasonal changes in the activity of soil micro-organisms. Cultures of *B. mutilate*, which actively break down BP, act similarly on a whole range of polycyclic aromatic hydrocarbons.

TABLE 7.13 *Mean contents of benzo(α)pyrene with standard deviations, in three crop species, as μg kg^{-1} (Ilnitsky et al., 1979)*

Crop	Initial concentration	Shoots	Roots
Wheat	1.58±0.08	19.73±2.79	8.44±2.70
Rye	0.74±0.07	8.01±0.61	1.60±0.10
Oats	1.27±0.06	14.30±0.78	5.05±0.38

In addition, bacteria isolated from sewage effluent, fresh-water systems and the sea have all been found to be capable of PAH degradation (Ermakov, 1979; Lembik, 1979; Shabad *et al.*, 1979).

4 Natural Origins of Polycyclic Aromatic Hydrocarbons

4.1 Synthesis by plants and micro-organisms

There has been some dispute as to whether or not plants and micro-organisms can synthesize PAH (Blumer, 1961; Gräf and Diehl, 1966). In order to exclude the possibility of exogenous PAH penetration, experiments were carried out in which a specially constructed, hermetically sealed chamber was used into which air was introduced after rigorous filtration. Every precaution was taken in germinating the seeds and in growing the resulting plants in the chamber to avoid contamination. The plants were harvested and subsequently analysed for BP content. The results given in Table 7.13 seem to provide evidence that BP is in fact formed.

4.2 Volcanic Activity

Volcanic activity represents a powerful pyrogenic process and the production of natural PAH is a likely outcome. Ilnitsky, Belitsky and Shabad (1975) collected samples of fresh volcanic ash, of soil and of plants near volcanoes, and analysed them for BP. The results are given in Table 7.14.

Calculations on the annual input of BP to the biosphere from volcanoes suggest that this amounts to only several tons. This appears to be an insignificant quantity in relation to man-made emissions which equal about 1300 tons annum^{-1} in the U.S.A. (Selkirk, 1977) and about 5000 tons globally each year (Suess, 1976).

TABLE 7.14 *Benzo(α)pyrene content (μg kg^{-1}) in soil, vegetation and volcanic mud in the vicinity of volcanoes (Ilnitsky, Belitsky and Shabad, 1975)*

Location	Soil			Vegetation			Volcanic mud		
	1	2	3	1	2	3	1	2	3
Burlyaschy Volcano	11	1.1	0.4–3.1	3	7.4	5.4–9.8	–	–	–
Uzon Volcano	5	4.2	0.9–7.1	10	3.6	0.7–11.7	12	2.8	0.5–7.5
Geyser Valley	5	1.3	0.3–0.4	4	4.7	1.1–8.4	–	–	–
Kliuchevskayasopka Volcano	13	3.4	0.4–14.6	6	7.4	1.4–16.4	–	–	–
Kurile islands: Kunashir	2	5.9	0.7–11.2	1	4.5	–	–	–	–

1 = no. of samples; 2 = mean content; 3 = confidence limits (p = 0.05)

5 Polycyclic Aromatic Hydrocarbons and Coal Conversion Processes

The large-scale introduction of coal liquefaction and gasification processes will lead to the formation of some products that are relatively rich in carcinogenic PAHs (Palmer, 1979). Special attention will have to be paid to this aspect of coal utilization, which is why considerable detail is given here of the "ambient" levels that might be expected from other natural and anthropogenic sources.

An increase in burning and chemical processing of coal might well give an increase in the potential hazards due to carcinogenic PAH. These would occur in the atmosphere and in products of coal processing such as residual tars and motor fuels where they might introduce new occupational hazards.

At the earliest stage of new or increased uses of coal the potential hazards should be recognized and plans made to control these hazards as cost-effectively as possible.

6 Polycyclic Aromatic Hydrocarbon Body Burdens in Man

It is now well established that the effect of chemical carcinogens is directly related both to the dose and to the duration of contact with the carcinogenic agent. It is thus important to know the "body load" to which man is exposed during his life-time. Man is exposed to PAH from the atmosphere, from food and from water. Over a 70-year life-span, food intake of PAH is likely to be about 5 mg. Inhabitants of large cities will, in addition, gain 12–16 mg from inhalation (Shabad and Dikun, 1959). Intake of PAH in water will normally be extremely low. Much of the PAH in water will be excreted, and Shabad and Dikun (1959) found that the tissues of elderly town-dwellers contained 1000 times less BP than might have been expected from intake figures. Linnik, Smirnov and Shabad (1971) showed that the activities of enzyme systems in the body lead to the formation of oxy-derivatives which are no longer carcinogenic in activity. Gorelova and Cherepanova (1970) confirmed this by carrying out experiments using cows, hens, ducks, geese and mice fed on food containing considerable quantities of BP. They emphasized the suitability for human consumption of meat, milk and even eggs obtained from these animals, even though the food that had been fed to these animals was contaminated. The PAHs appear in general to be more readily detoxified and excreted by a variety of species than are some of the halogenated pesticides and it appears unlikely, therefore, that PAHs will accumulate in food chains.

There has been considerable effort applied to the determination of maximum allowable concentrations (MAC) of PAH in the atmosphere since this represents the major exposure source to which man is subjected. BP levels are often taken as indicators for the establishment of MAC values and the standard can be set by reference to background levels that are continuously monitored. In addition, experiments utilizing animals that have been subjected to different routes of BP administration (dermal, oral, intra-tracheal) can give some useful indications. Thirdly, occupational exposures must be assessed. Standards set usually allow only 0.1 μg 100 m^{-3}. For higher figures, a 6-hr working day and period of extended leave are stipulated. Levels in water reservoirs are set at 0.005 $\mu g\ l^{-1}$ and levels in soil at 20 $\mu g\ kg^{-1}$.

It is not easy at this time to predict the levels of pollution by PAH that might result from increased coal combustion and the large-scale implementation of coal gasification and liquefaction operations. As these procedures become more widely used, monitoring of the working environment and of general environmental levels will have to be undertaken. The current standards can be taken as reasonable guidelines at the present time, but it must be borne in mind that any increase in dose and sustained contact may lead to increased disease and an increased incidence of malignant tumours. A comprehensive study involving adequate monitoring procedures will be required as a basis for the prophylactic measure that should accompany the increased utilization of coal.

7 References

ANDELMAN, J. B. and SUESS, M. J. (1970). Polynuclear aromatic hydrocarbons in the water environment. *Bulletin of the World Health Organisation*, **43**, 479–508.

BLUMER, M. (1961). Benzpyrene in soil. *Science*, **134**, 474–5.

BORNEFF, I. and FABIAN, B. (1963). Kanzerogene Substanzen in wasser und boden XIX. Weitere untersuchungen über polyzyklische aromatische kohlenwasserstoffe in erdproben. *Archiv für Hygiene und Bakteriologie*, **147**, 401–12.

BRASSER, L. J. (1979). PAH concentrations in the Netherlands. *Polycyclic Aromatic Hydrocarbons.* Colloquium Summaries. VDI, Kommission Reinhaltung der Luft.

BURLESON, G. R., CAULFIELD, M. J. and POLLARD, M. (1979). Ozonation of mutagenic and carcinogenic polyaromatic amines and polyaromatic hydrocarbons in water. *Cancer Research*, **39**, 2149–54.

COOK, J. W., HEWITT, C. L. and HIEGER, I. (1933). The isolation of a cancer producing hydrocarbon from coal-tar. *Journal of the Chemical Society*, **1**, 395–405.

Cook, J. W., Hieger, I., Kennaway, E. L. and Mayneord, W. V. (1932). The production of cancer by pure hydrocarbons. *Proceedings of the Royal Society,* **B111**, 455–84.

Egan, H., Castegnaro, M., Bogovski, P., Kunte, H., Walker, E. A. and Davis, W. (eds) (1979). *Environmental Carcinogens: Selected Methods of Analysis,* vol. 3, *Analysis of Polycyclic Aromatic Hydrocarbons in Environmental Samples.* IARC Publications no. 29. Lyon.

Ermakov, E. A. (1979). Certain data on the distribution of carcinogenic hydrocarbons in the bottom sediment, the surface layer of seas and the atmosphere. *Carcinogenic Substances in the Environment*, pp. 43–8. Gydrometeoizdat, Moscow.

Fox, A. S., Lindars, O.C. and Owen, P. A. (1974). Survey of occupational cancer in the rubber and cable making industries: results of five-year analysis, 1967–1971. *British Journal of Industrial Medicine,* **31**, 140–51.

Fritz, W., Engst, R., Shabad, Z. M., Dikun, P. P. and Khesina, A. Ya. (1977). Vergleichende Untersuchungen zur kontamination von lebensmitteln met krebserzengenden kohlenwasserstoffen in der UdSSR und der DDR. *Arch. Geschwulstforsch.*, **46/7**, 561–6.

Gofmekler, V. A., Manita, M. D., Manusadzhants, Zh. I. and Stepanov, Z. Z. (1963). The correlation between the concentration of 3,4,-benzpyrene and carbon monoxide in exhaust gases of automobiles. *Gygiena i Sanitariya*, **8**, 3–8.

Gorelova, N. D. and Cherepanova, A. I. (1970). On benzo(α)pyrene in cow's tissue and milk after its administration by fodder. *Voprosy Onkologii*, **16**, 69–73.

Gortalum, G. M. (1972). The circulation of polycyclic aromatic hydrocarbons in the marine environment. In *Investigations into the Prophylactic Treatment of Pollution of Man's Environment by Carcinogenic Substances,* pp. 30–32. Tallin.

Gräf, W. and Diehl, H. (1966). Über den naturbedingten normalpegel kenzerogener polyzyklischer aromatic und seine ursache. *Archiv fur Hygiene und Bakteriologie,* **150**, 49–59.

Grimmer, G. and Bohnke, H. (1979). Gas chromatographic profile analysis of polycyclic aromatic hydrocarbons in (I) high protein foods, (II) fats and vegetable oils and (III) plants soils and sewage sludge. In *Environmental Carcinogens: Selected Methods of Analysis,* vol. 3, *Analysis of Polycyclic Aromatic Hydrocarbons in Environmental Samples* (ed. by Egan, H., Castegnaro, M., Bogovski, P., Kunte, H., Walker, E. A. and Davis, W.). IARC Publications no. 29, pp. 63–173. Lyon.

Grover, P. L. (1980). The metabolic activation of the polycyclic aromatic hydrocarbons, *VDI Berichte*, **358**, 257–271.

Gurinov, B. P., Tugarinova, V. N. and Vasilieva, O. I. (1962). Carcinogenic properties of exhaust soot of motortransport. *Gygiena i Sanitariya,* **2**, 19–24.

Huenigen, E., Prietsch, V., Naumann, M., Jaskulla, N., Shabad, L. M., Khesina, A. Y. and Smirnov, G. A. (1978). Reduction of air pollution caused by carcinogenic substances from exhaust gases of automobiles. In *Atmospheric Protection*. The co-ordinating centre on the problems of atmospheric protection from pollution by harmful substances. Dresden, G.D.R.

Ilnitsky, A. P., Belitsky, G. A. and Shabad, L. M. (1975). Carcinogenic polycyclic aromatic hydrocarbon benze(α)pyrene in volcano exhausts. *Seriia Biologicheskaya, Dokladi Akademii Nauk USSR,* **255**, 214–16.

Ilnitsky, A. P., Khesina, A. Ya., Cherkinsky, S. N. and Shabad, L. M. (1968).

The influence of ozonation of the aromatic carcinogenic hydrocarbons. *Gygiena i Sanitariya*, **3**, 8–11.

Ilnitsky, A. P., Mischenko, V. S. and Shabad, L. M. (1979). The natural sources of carcinogenic hydrocarbons. In *Carcinogenic Substances in the Environment*, pp. 25–9. Gydrometeoizdat, Moscow.

Ilnitsky, A. P., Voronin, V. I. and Solenova, L. G. (1974). Carcinogenic effects of benzo(α)pyrene and its combination with other hydrocarbons during their oral intake. *Voprosy Onkologii,* **20** (10), 71–5.

Khesina, A., Ya. (1979). Determination of benzo(α)pyrene in extracts by spectroluminscence. In *Environmental Carcinogens: Selected Methods of Analysis,* vol. 3, *Analysis of Polycyclic Aromatic Hydrocarbons in Environmental Samples*, (ed. by Egan, H., Castegnaro, M., Bogovski, P., Kunte, H., Walker, E. A. and Davis, W.). IARC Publications no. 29, pp. 215–29. Lyon.

Klipinitser, I. L., Ninin, V. K., Kuznetsova, T. A., Shabad, L. M., Khesina, A. Y. and Serkovskaya, G. S. (1977). An investigation of benzo(α)pyrene in the products of the anthracene fraction of coal tar. *Gygiena Truda i Profzabolevanyi,* **3**, 47–9.

Konstantinov, V. G., Somakhina, P. G., Gotlieb, E. V. and Kuzminykh, A. I. (1974). On the question of carcinogenic danger in aluminium electrolysis workshops. In *Occupational Cancer*, pp. 87–91. Moscow.

Korfmacher, W. A. Natusch, D. F. S., Taylor, D. R., Mamantor, G. and Whery, E. L. (1980). Oxidative transformations of polycyclic aromatic hydrocarbons adsorbed on coal fly ash. *Science,* **207**, 763–5.

Lawther, P. J. and Waller, R. (1976). Coal fires, industrial emissions and motor vehicles as a source of environmental carcinogens. *Environmental Pollution and Carcinogenic Risks* pp. 27–40. (ed. by Rosenfeld, C. and Davis, W.). INSERM Symposium Series, 52, IARC Scientific Publications no. 13, Paris.

Lembik, J. L. (1979). On certain natural factors of benzo(α)pyrene destruction in fresh water reservoirs. *Carcinogenic Substances in the Environment*. Gydrometeoizdat, Moscow.

Linnik, A. B., Smirnov, G. A. and Shabad, L. M. (1971). The study of the carcinogenic activity of soot from aviation engines in experiments on animals. *Bulletin Experimentalnoy Biologhii i Meditsinii*, **2**, 83–7.

Mallet, L. and Héros, M. (1962). Pollution des terres vegetales par les hydrocarbures polybenzeniques due type benzo–3,3,-pyrene. *Compte Rendu Hebdomadaire des Séances de l'Académie des Sciences, Paris,* **245**, 958–60.

Palmer, A. (1979). Mortality experience of 50 workers with occupational exposures to the products of coal hydrogenation processes. *Journal of Occupational Medicine,* **21**, 41–4.

Petrov, N. I. (1961). *Manual of General Oncology*. Medghiz, Leningrad.

Pullman, A. and Pullman, B. (1955). Electronic structure and carcinogenic activity of aromatic molecules. *Advances in Cancer Research,* **3**, 117–69.

Selkirk, J. K. (1977). Benzo(α)pyrene cancero-genesis: a biochemical selection mechanism. *Journal of Toxicology and Environmental Health*, **2**, 1245–58.

Shabad, L. M. (1973). *Circulation of Carcinogens in the Environment.* Meditsina, Moscow.

Shabad, L. M. (1978). The International Symposium on air pollution and cancer in Stockholm, 1977. *Gygiena i Sanitariya,* **2**, 121–2.

Shabad, L. M. and Dikun, P. P. (1959). *Air Pollution by Carcinogenic Substances - 3,4,-Benzpyrene.* Medghiz, Leningrad.

SHABAD, L. M., KHESINA, A. YA. and KHITROVO, S. S. (1968). On carcinogenic hydrocarbons in the exhaust gases of motor car engines and their possible elimination. *Vestnik Akademii Meditsinski Nauk USSR*, **23** (6), 12–20.

SHABAD, L. M., KHESINA, A. YA., LINNIK, A. B. and SERKOVSKAYA, G. S. (1970). Possible carcinogenic hazards of several tars and of Locacorten-tar ointment: spectro-fluorescent investigations and experiments on animals. *International Journal of Cancer*, **6**, 314–8.

SHABAD, L. M., KHESINA, A. Y., SERVOSKAYA, G. S., OGLOBLINA, I. A. and FRIDMAN, Y. S. (1972). Determination of 3,5,-benzpyrene in protein fodder and in raw material employed during its obtention. In *Medicobiological Investigations of Hydrocarbon Yeasts* (ed. by Pokrovskii, A. A. Nauka, Moscow.

SHABAD, L. M., KHESINA, A. YA., TSIBAN, A. V., PANOV, G. V. and MIROSCHNICKENKO, N. M. (1979). Microbial degradation of carcinogenic polycyclic hydrocarbons in the marine environment. *Proceedings of the 14th Pacific Ocean Scientific Congress*, Khabarovsk.

SHABAD, L. M. and SMIRNOV, G. A. (1976). Aviation and environmental benzo-(α)pyrene pollution. *INSERM*, **52**, 53–60.

SIMS, P. and GROVER, P. L. (1974). Epoxides in polycyclic aromatic hydrocarbon metabolism and carcinogenesis. *Advances in Cancer Research*, **20**, 165–274.

SUESS, M. J. (1976). The environmental load and cycle of polycyclic aromatic hydrocarbons. *Science of the Total Environment*, **6**, 239–50.

SUESS, M. J. (1976). The environmental load and cycle of polycyclic hydrocarbons. *Proceedings of the International Conference on Environmental Sensing and Monitoring*, Las Vegas, 1975.

TAUSON, V. O. and VESELOV, I. (1973). On the bacterial degradation of cyclical compounds by sulphate reduction. *Mikrobiologiya*, **3**, 360–9.

TEUTSCHLAENDER, O. (1929). Uber den Pechkrebs der briket-arbeiter auf grund von fabrikseuchen in Baden und Sud Wales. *Zeitschrift Krebsforschung*, **28**, 283–300.

VARSHAVSKY, I. L., DOZENKO, V. I., IGNATOVICH, N. I. and KLYATSKIN, V. B. (1970). On methods of investigation of the content of carcinogenic hydrocarbons in the exhaust gases of autotransport. *Gygiena i Sanitariya*, **4**, 75–7.

WALLER, R. (1952). The benzpyrene content of town air. *British Journal of Cancer*, **6** (1), 8–21.

YAMAGIWA, K. and ICHIKAWA, K. (1918). The experimental study of the pathogenesis of carcinomas. *Cancer Research*, **3**, 1–29.

YANISHEVA, N. Y., CHERNICHENKO, I. A. and BALENKO, N. V. (1977). *Carcinogenic Substances and their Hygienic Control in the Environment*. Kiev.

YANISHEVA, N. YA. and KIREEVA, I. S. (1977). On the hygienic assessment of elements of the environment participating in the formation of the total dose of benzo(α)pyrene entering the human body. *Gygiena Naselënnoy Mestnosti*, **16**, 28–32.

YANISHEVA, N. YA. and KIREEVA, I. S. (1979). Determination of benzo(α)pyrene in air using quasi linear luminescence. In *Environmental Carcinogens: Selected Methods of Analysis*, vol. 3, *Analysis of Polycyclic Aromatic Hydrocarbons in Environmental Samples* (ed. by Egan, H., Castegnaro, M., Bogovski, P., Kunte, H., Walker, E. A. and Davis, W.). IARC Publications no. 29, pp. 63–173. Lyon.

YANISHEVA, N. YA. (1939). On the comprehensive approach to the study of benz(α)pyrene in the environment. In *Problemy Hygieny*, pp. 107–18 (ed. by

Gorski, T. and Pliss, G.). Society of Hygienists of Poland, Warsaw.

ZABEZHENSKY, M. A. (1968). City air pollution with carcinogenic substances released by motorcar engines. *Voprosy Onkologii,* **41** (1), 87–8.

ZDRAZIL, I. and PICHA, F. (1966). The occurrence of the carcinogenic compounds 3,4,-benzpyrene and arsenic in soil. *Neoplasma*, **13**, 49–55.

ZO BELL, C. E. (1969). Microbiological modification of crude oil in the sea. *Proceedings of the Conference on Prevention and Control of Oil Spills.* PAI Publications.

8

Potential Health Problems from Coal Conversion Technologies

1 Nature of the Problem

There are many economic and political pressures that lead, at present, to considerable scientific and technical activity in the field of coal liquefaction and gasification. It is important that within this framework the pollutant and health data available in published work should be assessed in order to identify the potential environmental and occupational hazards that may accompany these developments. As has been described in Chapter 4, the chemical structure of coal, the chemical changes that take place during its conversion, and the large quantities of coal that are likely to be utilized are the major concerns for evaluation of risks to health. These concerns are reinforced by the recent suggestion (Morris *et al.*, 1979) that energy inefficiencies, environmental effects and health damage increase in the order: (i) direct combustion of natural gas and oil; (ii) direct combustion of synthetic gas and oil; (iii) central-station electric power produced from synthetic gas; (iv) central-station electric power produced by the combustion of synthetic liquid fuels.

2 Areas of Concern

2.1 Available health data

Only one study of the actual health hazards of coal hydrogenation is presently available (Sexton, 1960a, 1960b; Ketcham and Norton, 1960; Weil and Condra, 1960). A group of 359 workers was examined for skin cancer over a period of 5 years (1955–60) at a liquefaction plant (West Virginia, U.S.A.) designed to process 273 t day^{-1} of incoming coal. Of this group, ten men developed skin cancer. Occupational exposure varied from several months to 23 years, although these longer exposures did not all accrue from work in the liquefaction plant, but all significant lesions were found in workers with less than 10 years' exposure. The reported incidence was 16–37 times the incidence of skin cancer expected in the chemical industry. Precancerous lesions were also found in the skin of 40 additional workers (Sexton, 1960b).

Analyses of the liquid materials produced in the plant identified a number of carcinogenic materials. An increase in carcinogenicity with increasing liquid boiling point, usually with respect to liquids with boiling points >260°C, was observed in laboratory animals (Weil and Condra, 1960). This relationship was particularly evident for the oil, produced in the plant, which was recycled to slurry the incoming coal. In addition, air measurements in working areas revealed benzo-(α)pyrene concentrations often in excess of 0.05 $\mu g\ m^{-3}$ with a peak value of 18 $\mu g\ m^{-3}$ (Ketcham and Norton, 1960). Benzo(α)pyrene has been identified as strongly carcinogenic (Chapter 7). These concentrations are in contrast to typical U.S. urban levels of 0.003–0.006 $\mu g\ m^{-3}$ (NAS, 1972), although higher concentrations are quoted for main roads in urban areas in some other countries (Shabad and Dikun, 1959; Shabad, Khesina and Khitrovo, 1968).

In 1977 a follow-up study was instituted to identify and trace the original 50 workers with skin lesions (Palmer, 1979). After a latency period of 18–20 years, the cancer incidence did not appear to be marked. It is well to note that none of the deaths had been autopsied (NIOSH, 1978a). It was not intended that the review by Palmer (1979) should be a basis for drawing firm conclusions about cancer incidence save for the highly selected group studied. An epidemiological study was recommended to identify the remainder of the work-force who were exposed to the coal hydrogenation plant discharges but who did not develop skin cancer in the 1955–60 interval. The possibility exists that some of the workers who had not developed skin lesions during that period may have developed more serious organ cancers after the surveillance programme was discontinued in 1960 (Palmer, 1979). It has also been reported that since the end of December 1958 the workers have been followed and that at first there were one or two new cases of skin cancer every year, but after the plant stopped operating the incidence was about one new case every 5 years (NIOSH, 1978b).

Mortality and morbidity data at other coal hydrogenation facilities have not been systematically collected or analysed, at least in the published literature. An inquiry survey of plants in the U.S.A.,

TABLE 8.1 *Relative risk of death for selected causes by length of employment at entry to study for coke-oven workers (Redmond et al., 1972).*

	5 years or more at coke oven	Less than 5 years at coke oven
Cause of death	Relative risk	Relative risk
All causes	1.08	0.83
Malignant neoplasms–lung, bronchus, and trachea	3.48	1.70
Malignant neoplasms–genitourinary organs	2.01	a
Other malignant neoplasms	0.93	0.40
Tuberculosis of the respiratory system	1.00	a
Other diseases of the respiratory system	1.28	0.88
Cardiovascular renal diseases	0.90	0.79
Accidents	0.88	1.12
All other causes	0.94	0.77

[a] Less than five deaths observed.

TABLE 8.2 *Relative risk of death for selected causes by coke oven subdivisions for coke oven workers 1951–66 (Redmond et al., 1972).*

Cause of death	Relative risk: Full topside	Partial topside	Side oven
All causes	1.12	1.07	0.92
Malignant neoplasms–lung, bronchus, and trachea	7.24	2.14	1.73
Malignant neoplasms–genitourinary system	a	a	2.02
Other malignant neoplasms	0.86	0.44	0.79
Tuberculosis of the respiratory system	a	a	1.06
Other diseases of the respiratory system	0.99	1.78	1.01
Cardiovascular renal diseases	0.84	1.18	0.80
Accidents	0.72	0.94	1.19
All other causes	1.00	0.76	0.85

[a] Less than 5 deaths observed.

South Africa, F.R.G., Scotland, and England by the U.S. National Institute for Occupational Safety and Health (NIOSH), uncovered no other reports of increased skin cancer incidence or other serious health effects in coal gasification or liquefaction plants (NIOSH, 1978b). However, there were no data quoted to support these negative observations. One death has been reported from a Koppers-Totzek plant in Modderfontein, South Africa. A laboratory assistant taking gas samples near the top of an electrostatic precipitator being purged with raw gas (approximately 58% CO) was overcome during a night-time temperature inversion (Partridge, 1976; Franzen and Goeke, 1976).

Health statistics on occupational groups in other coal conversion operations, such as coke ovens and coal-tar processing, have shown significantly higher lung cancer rates relative to groups without such occupational exposures (Doll *et al.*, 1965, 1972; Lloyd, 1971; Redmond *et al.*, 1972; Mazumdar *et al.*, 1975; Kawai, Amamoto and Harada, 1967). In particular, coke-oven workers appear to have 2.5 times the incidence of lung cancer of other steel workers (Lloyd, 1971). Tables 8.1–8.3 summarize some of these data, and Figure 8.1

TABLE 8.3 *Mean level of coal tar pitch volatiles (CTPV) exposure for coke-oven workers (Mazumdar et al., 1975)*

Exposure group	Average CTPV (mg m^{-3})
Topside	3.15
Side oven 1	1.99
Side oven 2	0.88

shows an estimated dose-response relationship based on the same information. British workers heavily exposed to the products of coal carbonization in gasworks, where coal is heated in retorts with the primary purpose of producing flammable gas, were found to have greater incidence of death due to lung cancer (140%) and bladder cancer (208%) compared to workers exposed only to by-products (Doll *et al.*, 1965, 1972). Analysis of gasworks air showed a general level of about 3 μg m^{-3} of benzo(α)pyrene and over 200 μg m^{-3} in the worst working area (Lawther, Commins and Waller, 1965).

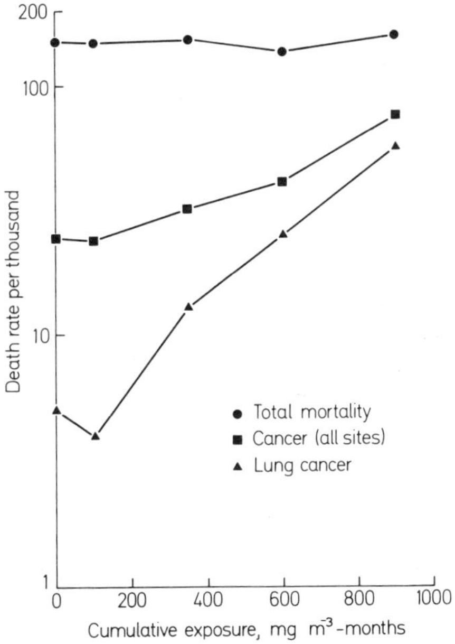

FIGURE 8.1 *Age-adjusted cancer death rates among non-white coke-oven workers by cumulative coal tar pitch volatile exposure, 1951–66 (Mazumdar et al., 1975)*

The environment on top of a coke oven or in a gasworks is, of course, different from that likely to occur in hydrogenation plants which will eventually be designed to process raw materials continuously. Continuous processing requires more complete containment (and consequently better control) of all substances than do batch operations such as coke ovens.

2.2 *Environmental monitoring data*

A considerable number of full-scale plants have been built for thermo-chemical conversion of coal. Tables 4.3 and 4.4 have listed some of these installations. However, although design values are available for such major components as steam and coal feed rates and cooling water requirements (Ghassemi, Crawford and Quinlivan, 1978; Parker and Dykstra, 1978), until recently very little actual operating information had been published. Consequently, judgements about potential environmental discharges were often based on bench-scale or pilot plant experimental runs. These data are useful qualitatively, but probably will not quantitatively (or sometimes even qualitatively) mirror the performance of a full-scale plant. Pilot plants are usually operated to test a great variety of operating conditions. Steady-state operation is attained at most for a few days. On the other hand, commercial plants operate continuously for weeks or months and many pollution problems will become evident only after this time. The differences in equipment size between research and full-scale processes may also result in alteration of the characteristics of discharge pollutants.

Only limited data have been published for air or water concentrations in the vicinity of operating plants which might mirror community exposures. The most severe potential hazard to exposed populations may occur when control equipment designed for continuous operation fails to function efficiently during start-up, shut-down and upset periods. An example of this is the reported formation of nickel carbonyl ($Ni(CO)_4$), in methanizer gas in plants producing high Btu product gas (Ghassemi, Crawford and Quinlivan, 1979). A nickel compound is used to catalyse the conversion of shift gas to product gas. Since temperatures >275°C cause increased conversion of CO to CH_4 as well as discouraging the formation of $Ni(CO)_4$, continuous processing temperatures are well above this level. However, during start-up and shut-down, cooler temperatures are encountered which require pressurizing the methanizer equipment with an inert gas (e.g. N_2) until proper

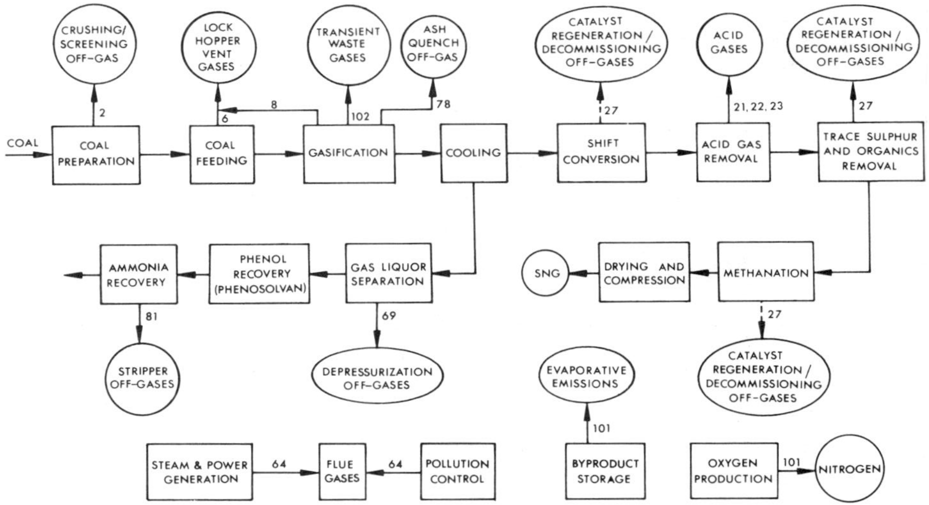

Stream	Identification	Stream	Identification
21,22,23	Combined gases from Rectisol unit	81	Stripper off-gas
2	Fugitive coal dust	69	Depressurization gas
6	Coal lockhopper vent gas	66	Nitrogen
8	Ash lockhopper vent gas	64	Utility boiler flue gas
102	Transient waste gases	101	By-product storage vent gases
78	Ash quench off-gas	27	Catalyst decommissioning off-gas
27	Catalyst decommissioning off-gas		

FIGURE 8.2 *Process modules generating gaseous wastes in Lurgi synthetic natural gas (SNG) systems (Ghassemi, Crawford and Quinlivan, 1979)*

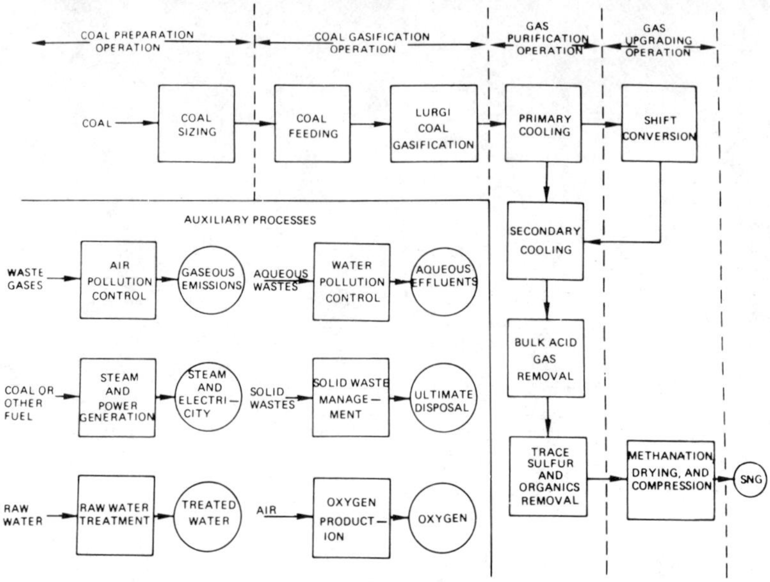

FIGURE 8.3 *Generalized process flow diagram for Lurgi systems producing synthetic natural gas (SNG) (Ghassemi, Crawford and Quinlivan, 1979)*

operating conditions are achieved. Additionally the catalyst must be periodically reactivated by burning off deposited sulphur and carbon compounds. The waste gas from this operation is likely to contain CO and CS as well as $Ni(CO)_4$ and must be carefully monitored and controlled.

2.3 Types of hazardous discharges

2.3.1 Steady-state operation

Representative examples of the various waste streams which need to be controlled are given in Figures 8.2–8.5 for the Lurgi process, recognizing that such an analysis will vary from process to process. The flow sheets do not describe additional wastes which may result from pollution control operations, e.g., solid $CaSO_4 \cdot 2H_2O$ from SO_2 scrubbing. Many of the indicated streams have not been

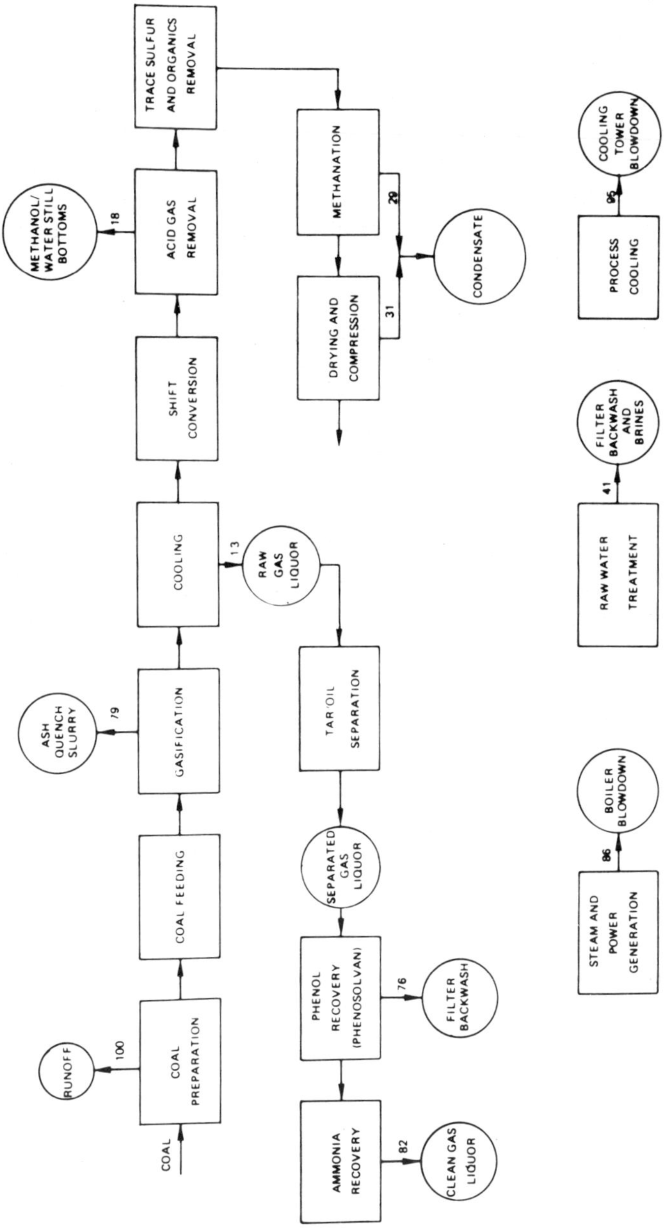

FIGURE 8.4 *Process modules generating aqueous wastes in an integrated Lurgi synthetic natural gas (SNG) system (Ghassemi, Crawford and Quinlivan, 1979)*

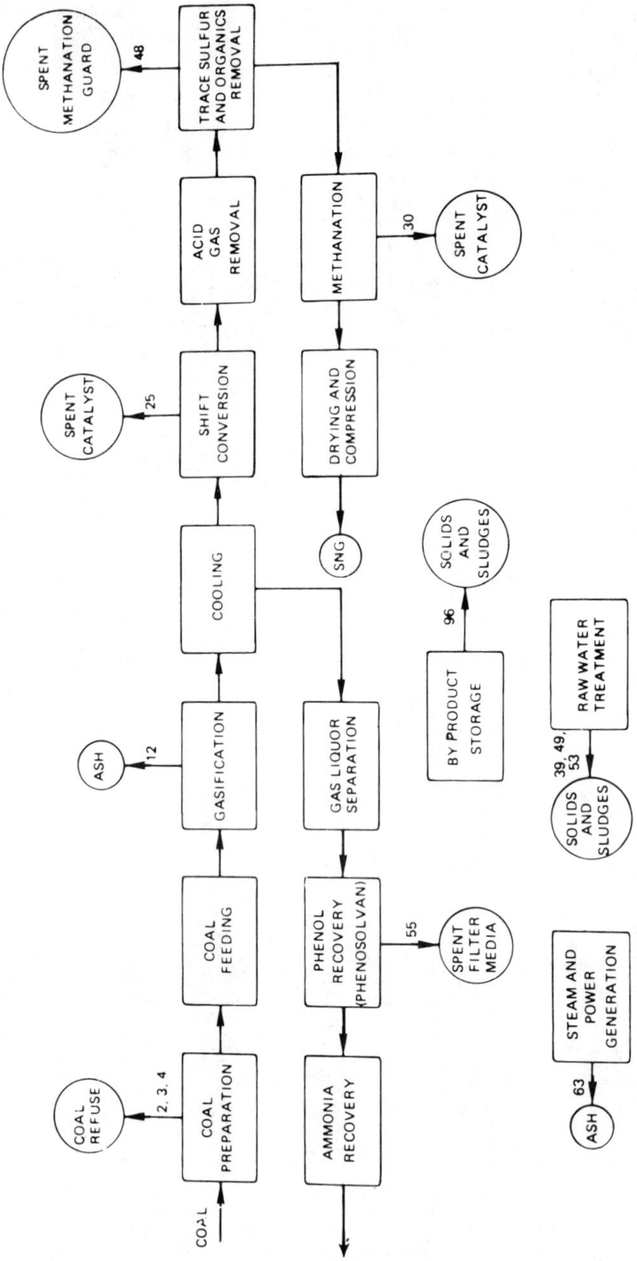

FIGURE 8.5 *Process modules generating solid wastes in Lurgi synthetic natural gas (SNG) systems (Ghassemi, Crawford and Quinlivan, 1979)*

TABLE 8.4 *Lurgi gasification product gas (Ghassemi, Crawford and Quinlivan, 1979).*

	Coal no.[a]				
Type/Origin	1[b] Montana Rosebud sub-bituminous	2[b] Illinois 6 high volatile	3[b] Illiniois 5 bituminous	4[b] Pittsburgh 8 bituminous	5[c] South African sub-bituminous
Gas production rate – $Nm^3\ kg^{-1}$ (scf $1b^{-1}$ MAF[d])	2.0 (34)	2.3 (39)	2.3 (39)	2.4 (41)	2.3 (39)
Gas composition (mole %) dry					
H_2	41.1	39.1	38.8	39.4	40.05
O_2	1.2	0.6	0.8	0.8	–
N_2 + Ar	–	0.6	0.7	0.8	–
CO_2	30.4	31.2	31.0	31.5	28.78
CO	15.1	17.3	17.5	16.9	20.20
CH_4	11.2	9.4	9.2	9.0	8.84
C_2H_4	–	–	–	–	–
C_2H_6	0.50	0.70	0.50	0.70	0.54
H_2S	0.50	1.1	1.1	0.8	–
Gas composition (mole %) dry					
COS, CS_2, Mercaptans, Thiophenes	315 ppm	180 ppm	232 ppm	122 ppm	–
CS2 (South African)					–
Mercaptans (South African)					–
Thiophenes (South African)					–
HCN	0.0002	0.0023	0.0078	0.00066	–
NH_3	0.00012	0.0002	–	0.0002	–

[a] Data for coals 1, 2, and 4 represent gas after cooling and moisture condensation such that some tar, oil, NH_3 and particles have already been removed.
[b] Data from Westfield, Scotland gasifier. [c] Data from Sasol, South Africa. [d] Moisture and ash free basis.

TABLE 8.5 *Average composition of Wellman-Galusha product gas*[a] *(Thomas, Trede and Page, 1979)*

Component	Volume % (dry)
CO_2	5.5 (5.0–7.5)[b]
H_2	16.3
O_2	0.9
N_2	51.6
CH_4	0.2
CO	25.5 (25–26)[b]
NH_3	180 (100–200)[b] ppm
CN^-	32 ppm
SCN^-	8 ppm
C_1	1910 (1500–4500)[b] ppm
C_2	< 1 ppm
C_3	3 ppm
$Fe(CO)_5$	0.004 ppm
$Ni(CO)_4$	0.01 ppm
H_2S	690 (600–700)[b] ppm
COS	93 (70–100)[b] ppm
SO_2	21 (4– 30)[b] ppm
CS_2	< 1 (< 10)[b] ppm
Total sulphur	730 ppm
Heating value (dry, @ 70°F)	5.1 MJ m^{-3} 137 Btu scf^{-1}

[a] Nominal 21.8 t day^{-1}.
[b] Ranges for on-line gas chromatograph results.

well-characterized, particularly for small fractions of polycyclic hydrocarbon material, sulphur compounds and trace metals.

Certain discharges from coal hydrogenation processes are unusual compared to those from ordinary chemical plants. The following discussion gives examples of such process streams which may enter the environment. Unless otherwise noted, the data were taken from pilot plants processing >10 t of coal per day or from full-scale installations. Table 8.4 shows the composition of gas from a Lurgi gasifier. This stream is perhaps representative of the transient waste gas stream (stream 102) in Figure 8.2 and also reflects the types of materials that must be removed later in the process. The presence of HCN and a variety of sulphur compounds, as well as elevated CO concentrations, is notable. The same observations can be made about low-Btu product gas from a Wellman-Galusha gasifier (Table 8.5). However, more detailed analysis of this stream also revealed significant amounts of $Ni(CO)_4$ and iron carbonyl ($Fe(CO)_5$) (Thomas, Trede and Page, 1979).

The acid gases (streams 21, 22 and 23) in Figure 8.2 will be a major discharge volume. Table 8.6 lists some of the constituents. Significant among these are the very high concentrations of CO, H_2S and CO_2. The public health implications of the latter are not well understood, but a typical gasification plant will emit in the neighbourhood of 16 000 t day^{-1} (for health effects review see NIOSH, 1976b).

Experience in Czechoslovakia (Mácak, Prášek and Šimánek, 1976) suggests that high-pressure gasification of high-sulphur coal results in 81–87% of the sulphur passing into the crude gas as hydrogen sulphide and 10% being bound in the ash. Hence a relatively low sulphur tar is formed (av. 1% by wt). The H_2S can be readily converted to sulphur, sulphuric acid or carbon disulphide and economically utilized.

The high sulphur content of brown coal is often accompanied by a high arsenic content ($\simeq$100 ppm As). This accumulates mainly in the ash and tar, 51.7–57.4% and 24.2–25% respectively, with lesser amounts being emitted in crude gas as arsenic trihydride (0.1–0.2 mg m^{-3}). The transfer of arsenic into gaseous and liquid forms is important both in view of the element's toxicity and of the effect it has on the final treatment of waste water and gases (Prášek, 1980).

A recent study of the Lurgi Kosovo Kombinant plant in Pristina, Yugoslavia represents the most complete stream characterization data-set for a commercial coal gasification plant in the published literature at the present time (Bombaugh *et al.*, 1980). Although more data will be forthcoming, the information reported here supplies a representative evaluation of a plant of Lurgi design, at steady state, with relatively few environmental controls. Table 8.7 summarizes "best value" concentrations of vented gas streams (3.6, 13.6, and 7.1 being subsequently sent to flare) based on repeated sampling, engineering and analytical judgement. Streams 7.1 (H_2S vent) and 7.2 (CO_2 vent) represent discharges from the Rectisol acid gas clean-up step (streams 21, 22 and 23 in Figure 8.2) and would ordinarily be processed further for H_2S removal. Even so, the concentrations of CO, HCN, NH_3 and various mercaptans (CH_3SH, C_2H_5SH) are significant in most of the streams, and the

TABLE 8.6 *Characteristics of acid gases produced by the Rectisol process (streams 21, 22, 23 of Figure 8.2) (Ghassemi, Crawford and Quinlivan, 1979).*

Process	1	2			3			4			5		
	Type A[a]	Type A[a,b]			Type B[a]			Type B[a]			Type B[c]		
Constituents/ Parameters	22	21	22	23	21	22	23	21	22	23	21	22	23
H_2	0.4	21.4	2.6	0.14	0.15	0.79	–	0.76	–	–	0.33	–	–
CO	0.014	18.2	4.8	0.0	0.05	0.22	–	0.11	–	–	0.14	–	–
CH_4	0.017	11.4	7.2	0.9	–	0.05	–	0.06	–	–	0.00	–	–
CO_2	73.95	46.7	83.4	97.2	76.81	98.91	64.6	90.85	–	68.31	80.19	–	68.46
N_2 + Ar	25.62	1.5	0.8	0.03	23.0[d]	0.05	0.1	8.22	–	1.92	19.34	–	–
H_2S	–	3176 ppm	4941 ppm	8824 ppm	2 ppm	2 ppm	35.2	5 ppm	–	29.77	<5 ppm	–	30.78
COS	–	–	–	0.003	–	–	0.1	–	–	–	8 ppm	–	0.76
C_2^+	–	0.7	1.1	0.7	–	–	–	–	–	–	–	–	–
MeOH	–	–	–	–	–	–	–	–	–	–	–	–	–
CS_2	–	–	–	0.0002	–	–	–	–	–	–	–	–	–
RSH	–	–	–	0.028	–		–	–	–	–	–	–	
Thiophene	–	–	–	0.0002	–	–	–	–	–	–	–	–	–
Temp:°K(°F)	–	273(32)	273(32)	268(23)	–	–	–	–	–	–	295(72)	–	322(121)
Pressure: MPa (psia)	0.1(15)	1.3(195)	0.46(70)	0.1(15)	0.1(15)	0.24(36)	0.24(36)	0.1(15)	–	0.2(28)	0.1(15)	–	0.5(73)
Rate: $Nm^3 hr^{-1}$	45 090	4500	15 000	98 000	41 480	14 130	1980	50 280	–	2390	30 800	–	673
(scf min^{-1})	(27 956)	(2852)	(9300)	(60 760)	(25 845)	(8760)	(1230)	(21 170)	–	(1480)	(19 100)	–	(417)

[a] Type A – simultaneous removal of CO_2 and H_2S with simultaneous recovery of CO_2 and H_2S.
Type B – simultaneous removal of CO_2 and H_2S with separate recovery of CO_2 and H_2S.
[b] Data are for Sasol, South Africa Lurgi plant.
[c] Data are for an oil gasification plant using the Texaco gasification process.
[d] Includes N_2 stripper at 358 000 $Nm^3 hr^{-1}$ (223 000 scf min^{-1}).

TABLE 8.7 *Vented gas streams for Kosovo Lurgi gasification process*[a] *(Bombaugh et al., 1980)*

	Stream								
	3.2 Lockhopper vent gases	3.6	13.1	13.3	13.6	13.7	14.5	7.1	7.2
	low pressure	high pressure	Tar tank vent	Medium oil tank vent	Tar separation expansion gases	Phenolic H_2O tank vent	Stripper vent	H_2S vent	CO_2 vent
Fixed gases (vol.%)									
H_2	37.0	32.0	Trace	Trace	11.0	Trace	Not found	0.7	Trace
O_2	0.27	0.24	19.0	0.45	Trace	13.0	–	Trace	Trace
N_2	0.18	0.14	77.5	1.1	Trace	39.0	–	Trace	Trace
CH_4	8.6	10.5	0.16	7.6	3.5	0.2	Trace	4.3	1.2
CO	14.6	12.0	Trace	5.9	1.1	Not found	Not found	1.1	Trace
CO_2	36.5	42.0	0.86	69.0	77.5	35.0	55.0	88.0	94.0
Hydrocarbons (vol.%)									
C_2–C_5	0.41	0.79	0.01	0.98	1.24	0.066	Trace	1.81	1.88
C_6+	0.12	0.08	0.37	2.4	1.3	1.8	Not found	0.21	Not found
Aromatic hydrocarbons (ppmv)									
Benzene	760	550	2000	7650	9600	11 000	–	110	1.0
Toluene	220	100	960	1400	1200	2300	–	8	Trace
Xylene & Ethylbenzene	75	38	220	140	1500	280	–	–	1.4
Phenols	6	2	22	45	4.2	Trace	6200	Not found	Not found
Other compounds (ppmv)									
H_2S	12 700	3500	6900	26 000	9000	12 600	19 500	45 400	39
COS	110	120	110	96	120	41	Not found	420	62
CH_3SH	420	460	390	5200	2500	2100	290	2100	8.5
C_2H_5SH	220	210	240	2100	1600	7200	100	780	4.4
NH_3	2400	Not found	2600	11.4	19 300	12 000	418 000	2200	46
HCN	600	170	130	–	65	38	4800	200	13
Flow rates[b], Nm^3 gasifier-hr^{-1} at 25°C	21	230	0.51	1.7	28	5.5	260	3600	3600

[a] Nominal 384 t day^{-1} coal feed/gasifier.
[b] Ordinarily five or six Kosovo gasifiers are in operation at one time.

TABLE 8.8 *Polycyclic aromatic hydrocarbons in Kosovo product and aqueous streams (Bombaugh et al., 1980)*

	Phenolic water (mg l^{-1})	Phenosolvan waste-water (stream 76, Figure 8.4) (mg l^{-1})	Light tar (mg kg^{-1})	Medium oil (mg kg^{-1})
Benzo(α)anthracene	0.92	Not found	490	156
7,12-dimethylbenz(α)anthracene	0.22	Not found	1090	62
Benzo(β)flouroanthrene	0.68	Not found	306	115
Benzo(α)pyrene	0.18	Not found	210	68
3-methylcholanthrene	<0.004	Not found	26	Not found
Dibenz(α,h)anthracene	0.02	Not found	23	6.6
252 Group (as BP)	1.26	0.19	945	282

quantities emitted are notable. Also of considerable concern are the relatively high concentrations of benzene and other aromatics.

Table 8.8 shows the polycyclic aromatic hydrocarbon content of the Kosovo oil, tar and phenolic water. Using these data and the oil and tar particulate fractions in waste gas streams, the benzo(α)pyrene content of flare and other exhaust gases has been estimated in Table 8.9. It is evident that if these estimates prove to be accurate, the discharge of benzo(α)pyrene is significant. Analyses of these streams for polycyclics are presently being carried out.

The Kosovo stream characterizations are very useful, as they can be used in conjunction with measured flow rates, appropriate air quality models and meteorological conditions to predict human exposure. Methods are also available for estimating some of the health effects which might result from such designs (e.g. Wadden, 1978).

An ambient monitoring study was also carried out in conjunction with the Kosovo process characterization (Patterson, 1980). Measurements of gas composition, particulate matter and meteorological variables were collected at four locations approximately 2 km from the plant complex. When the data become available, they should provide a valuable experimental link between the emission streams and likely human exposures.

Many liquid streams are likely to contain phenols, cyanide and suspended tar and oil. The Sasol clean gas liquor (stream 82, Figure 8.4 and Table 8.10) is a representative example. Although the Lurgi process does not require recycle oil for forming a coal slurry for

TABLE 8.9 *Estimated emissions of benzo(α)pyrene from the Kosovo Lurgi gasification process based on analysis of tars and oils (Bombaugh et al., 1980)*

	Stream No.			
	3.2 Low pressure lockhopper vent	3.6 High pressure lockhopper vent	3.6 Tar separation waste gas stream	20.1 Combined gas to flare
Dry gas flow rate (m^3 gasifier-hr^{-1} at 25°C)	21	230	28	1330
Total particulate (mg m^{-3} at 25°C)	8100	960	920	380
Tars and oils (mg m^{-3} at 25°C)	7300	660	660	290
BP based on tar (210 mg kg^{-1}) (mg m^{-3} at 25°C)	1.5	0.14	0.14	0.061
BP based on oil (68 mg kg^{-1}) (mg m^{-3} at 25°C)	0.50	0.044	0.044	0.020

subsequent gasification, many liquefaction processes do. This is of concern in the light of the skin cancer incidence in the West Virginia plant and the carcinogenicity of the recycle oils in the process (Sexton, 1960a, 1960b; Ketcham and Norton, 1960; Weil and Condra, 1960). Table 8.11 (Lurgi product oil) and Table 8.12 (solvent-refined coal (SRC) recycle solvent) emphasize the highly aromatic content of oil produced in these processes. Table 8.13 (Synthoil) gives examples of the benzo(α)pyrene concentration in oils from a bench-scale pilot liquefaction reactor. Since many polycyclic aromatic hydrocarbons have been identified as hazardous to man, the potential of these liquids for causing adverse effects on the central nervous system, kidneys and skin, as well as their possible carcinogenicity, needs a continued analysis.

Table 8.14 shows a comparison between a number of trace elements in coal and the corresponding ash from several Lurgi gasifiers. Potentially hazardous substances such as As, B, Be, Cr, Cu, Mn, Ni, Pb, Sb, and V appear to concentrate in the ash while volatile components such as Hg leave in the gas phase. What is not included in this analysis is the hydrocarbon material associated with

TABLE 8.10 *Properties of clean gas liquor at the Sasol phenolsolvan plant*

Parameter/constituent	Clean gas liquor	
	a	b
Total phenols	160	–
Steam volatile phenols	1	–
COD	1126	1330
Fatty acids (as CH_3COOH)	560	–
Total suspended solids	21	–
Total dissolved solids	875	596
Suspended tar and oil	21	–
Total ammonia	215	150
Total sulphides	12	<0.5
Cyanide	1	–
Thiocyanate	–	80
Fluoride	56	–
Chloride	25	–
Sodium	–	–
Calcium	18	–
Iron	1	–
Orthophosphate	2.5	0.2
Conductivity ($\mu S\ cm^{-1}$)	1000–1800	–
Total organic carbon	–	–
pH	8.4	8.2

All units in $mg\ l^{-1}$ except for pH and electrical conductivity.
[a] Gluskoter *et al.* (1977).
[b] Cameron (1974).

the ash. Some phenolic material is likely to be disposed of with the waste ash and char, and has the potential to be leached out in water systems (Ketcham and Norton, 1960). For instance, the slags from a 13.6 t day^{-1} Texaco coal gasification process operating on western coals contained 18–230 ppb phenanthrene (including anthracene) and 10–21 ppb acenaphthene, as well as other cyclic hydrocarbon compounds (Schlinger and Richter, 1980). Other important solid wastes are spent catalysts (streams 25 and 30, Figure 8.5) which will contain such metals as Ni, Co, and Mo.

Several mutagens including thioacetamide, acetanilide, hydroxylamine and *n*-propylmethane sulphonate already have been identified from coal conversion processes (Brown and Witter, 1979). Both the

TABLE 8.11 *Organic composition of Lurgi oil produced at the Westfield, Scotland facility (Ghassemi, Crawford and Quinlivan, 1979)*

Compound/class	Concentration (wt%)
Paraffins	10.71
Aromatics	89.30
Percentage aromatics as:	
Benzene	19.56
Toluene	28.40
Xylene and ethyl benzene	14.70
Ethyl toluene	2.69
Trimethyl benzenes	11.80
Styrene	1.07
Indane	1.43
1,2-benzofuran	1.09
Indene	5.37
Naphthalene	1.40
Thiophenes	1.77

strength of their mutagenicity and the locations in the process where they are produced need to be determined.

Other studies (Antizzo, 1978) have stressed the problems resulting from the impossibility of recycling all the waste-waters produced and also the magnitude of the solid wastes generated (Thode and Crowther, 1980) by a single large-scale coal conversion facility (20–40% of the mass of coal feed to the processes). Concern for the potential impact regarding air quality, water quality and ecosystems, as well as human health, has been expressed in a number of environmental development plans (Almaula and Nardella, 1978; Hemenway and Almaula, 1978).

2.3.2 Non-steady-state operation

While relatively few data are available for steady-state operating conditions, even less exists for start-up, shut-down and upset conditions.

The following information, however, is available from the experience of start-up operations of pressurized gas plants in Czechoslovakia. Because of safety considerations the first gas under pressure that is produced cannot be utilized and this is usually emitted to the atmosphere. From this emission something in the order of 7.6, 400

TABLE 8.12 *Composition of solvent refined coal (SRC) recycle solvent*[a] *(Schiller, 1977)*

Probable structural type	Sample SRC 115R (wt %)	Sample SRC 308[b] (wt %)	Sample SRC 122[c] (wt %)
Tetralin	2.84	–	–
Tetrahydroacenaphthene/Indene	1.35	0.22	0.24
Naphthalene	7.42	14.27	11.10
Acenaphthene/Biphenyl	18.73	12.57	21.89
Fluorine/Acenaphthalene	9.95	7.60	6.71
Phenanthrene/Anthracene	7.61	16.55	14.51
Dihydropyrene	5.25	7.00	5.99
Pyrene/Fluoranthene	6.95	4.49	3.56
Phenylpyridine/Tetrahydroacridine	0.75	3.00	1.74
Carbazole	1.87	2.55	1.77
Phenols	10.73	0.13	1.82
Indanols	2.89	0.88	1.76
Dibenzolfuran	6.48	9.87	9.10
Benzofuran	1.75	1.74	–
Hydroxyacenaphthene	2.84	–	2.68
Hydroxyanthracene	0.20	–	–
Paraffins	4.22	7.14	6.19
Monocycloparaffins	5.92	3.84	3.98
Dicycloparaffins	2.24	1.39	1.62
Acridine	–	4.17	2.62

[a]Nominal 45.4 t day^{-1} coal feed.
[b]Also present 0.42% chrysene, 1.36% quinoline, 0.09% naphthenobenzoquinolines, 0.69% azapyrene/benzocarbazole.
[c]Also present 0.83% quinoline, 0.20% azapyrene/benzocarbazole, 0.91% benzonaphthofuran.

TABLE 8.13 *Polycyclic aromatic hydrocarbons in Synthoil*[a] *(Schultz, 1976)*

Compound	Concentration (ppm)
Phenanthrene	413
Benzo(α)anthracene	18
Benzo(α)pyrene	41

[a]Nominal 0.5 t day^{-1} coal feed.

TABLE 8.14 *Coal and ash trace element composition (ppm) in Lurgi gasifiers (adapted from Ghassemi, Crawford and Quinlivan, 1979)*

	Sub-bituminous Montana/Rosebud[a]		Bituminous Illinois 6[a]		Bituminous Illinois 6[c]		Bituminous Illinois 5[a]		Bituminous South Africa[b]	
Element	Coal	Ash	Coal	Ash	Coal	Ash	Coal	Ash	Coal	Ash
As	0.08–1.2	26	1	0.1	1	3	2	0.3	2–5	1–2
B	32	380	132	622	132	355	307	673	100	100–150
Be	0.7–0.8	2.8	1.6	14	1.6	12	2	20	2–3	0.05
Cd	0.31–0.8	2.4	<0.4	<0.3	<0.4	<1.6	–	0.3	<0.1	<0.1
Co	0.6–4	4.3	4	0.4	4	34	4	0.4	–	–
Cr	4–16	440	20	750	20	212	15	592	–	–
Cu	9–10	130	12	239	12	57	–	273	–	–
F	66	600	79	5	79	<10	57	5	100	150
Hg	0.17–11	0.03	1.1	0.04	1.1	0.05	0.2	0.016	<0.1	–
Mo	2.2	200	7	6	7	30	7	8	–	–
Mn	2.8–3.4	790	20	200	20	1859	0.22	305	500	2000
Ni	2–14	200	14	456	14	89	32	452	30–50	150–200
Pb	0.51–12	40	10	96	10	45	28	200	10–20	50
Sb	–	6.2	0.1	0.2	0.1	4.2	0.2	19	<0.5	<0.5
Se	0.33	2.2	1.3	–	1.3	<1	1.3	–	0.3	–
Sn	0.14	3.7	–	–	–	–	9	–	–	–
V	10–14	91	20	301	20	184	21	181	300–500	1000

[a]From tests at Westfield, Scotland.
[b]From Sasol, South Africa.
[c]Illinois State Geological Survey.

and 800 kg of H_2S, CO and CO_2 respectively are released for every 1 $\times 10^6$ m^3 of gas eventually generated. Tar and phenol are also emitted in a vaporized form. Proposals for reducing such emissions include a shortening of the start-up time and the initial use of large quantities of pressurized air (Prášek, 1980).

There are also several other concerns about health and safety hazards that should be kept in mind (NIOSH, 1978a):

(1) Inert atmospheres: oxygen concentrations well below 19% can be expected in all process vessels and these should be considered hazardous areas.
(2) Explosive atmospheres: enclosed areas exposed to hydrogenation products or by-products may contain gases or vapours from volatile liquids capable of forming explosive mixtures in the air. In coal preparation areas, explosive concentrations of coal dust may be present in enclosed spaces.
(3) Surface contamination: vessel surfaces exposed to products or by-products of hydrogenation will be contaminated, in some cases, by condensed coal gasification products containing carcinogenic material. For instance, an examination of a pilot plant with ultraviolet light showed extensive surface contamination which probably contained carcinogenic hydrocarbons (Bolton, 1976).

2.3.3 Accidents

The types of accidents which may occur in hydrogenation plants are perhaps not significantly different from those in other installations processing flammable materials. For instance, Partridge (1976) reports that a Koppers–Totzek plant (Modderfontein, South Africa, Table 4.4) has experienced several explosions (in a coal dust bunker and in a washtower downstream of the gasifier). However, in contrast with more conventional processes, the materials emitted may well be quite hazardous. As an example, a methanator fire could well release $Ni(CO)_4$ and considerable amounts of CO and HCN.

Although not in the major accident category, leaks have also been a problem at the South African plant. This experience is probably not unique. Numerous shut-downs during the first months of start-up operation resulted in considerable thermal cycling. This in turn led to many leaks from pipe and vessel joints, particularly in the boiler feed water and steam systems and from process piping and vessels in the CO shift unit (Partridge, 1976). Leaks also appeared to be significant at the West Virginia pilot plant (Ketcham and Norton, 1960).

Another type of "leak", pollutant escape through fissures in collapsed overburden, has been reported for *in situ* gasification in Russia (C&EN, 1979).

2.4 Possible Exposed Populations

2.4.1 Workers and dependants

This group represents those who are most likely to be exposed to the highest concentration of hazardous materials. Dependants are at risk to workplace hazards brought into the home on clothes and skin.

NIOSH (1978a) has noted that significant toxic hazards include polynuclear aromatic hydrocarbons (Tables 8.8, 8.9, and 8.13) and other carcinogens such as β-naphthylamine and benzidine, nitrosoamines and $Ni(CO)_4$ (Table 8.5). Compounds capable of potentiating carcinogenesis such as SO_2, acridines, dodecane and phenols are also present. Organs conspicuously at risk are the skin and lung, other parts of the respiratory tract and the genitourinary system, especially the bladder. Effects other than cancer, such as dermatitis of the skin and liver, and central nervous system (CNS) disorders from phenols, H_2S, COS, CS_2, iron and nickel carbonyl (Tables 8.5 and 8.7) and aromatic amines, may be significant. Non-maligant respiratory effects, such as chronic bronchitis, as well as cardiovascular effects due to CO and CS_2, are also possible problems. Acute lung irritation can be caused by a wide variety of pollutants such as HCN, NH_3, SO_2, H_2S, and phenols (Tables 8.5 and 8.7). Benzene (see Tables 8.7 and 8.11) is myelotoxic and leukaemogenic (NIOSH, 1976a).

Several groups of workers may be at high risk to coal hydrogenation emissions. Workers exposed for more than 5 years or who are over 45 are possibly more susceptible to skin cancer (Table 8.1). If there are operations involving high exposure, workers concerned with these will be at risk. Appearance of skin cancer may be a warning sign for subsequent cancer in the respiratory and digestive tracts. Holmes, Kipling and Waterhouse (1970) reported a significant association of these effects in men with scrotal cancer.

2.4.2 Communities in the plant vicinity

Both normal and hypersusceptible populations need to be considered in evaluating community exposure. Groups hypersusceptible to hydrogenation discharges include children, the elderly and those with illness or genetic dysfunction which would make them more vulnerable to the stress of exposure.

Collection and periodic review of cancer, disease and birth defect registries are necessary requirements of community health and surveillance. Monitoring for pollutants in air and water in the plan vicinity would also be prudent. When pollutant quantities discharged from the plant can be determined, methods are available for approximating community exposures. Considerable research effort is required before the exposure commitment methodology can be applied to such situations in practice.

2.4.3 Product users

The chemical composition of liquid and gaseous products from coal hydrogenation is different from the composition of equivalent materials extracted from petroleum or natural gas. Product oils are likely to have high aromatic contents (Table 8.11) and significant amounts of polynuclear hydrocarbons, some of which are carcinogenic (Tables 8.8 and 8.13). It has been reported that Russian data indicate that gasoline from coal produces skin rashes while gasoline from petroleum does not produce a rash under similar circumstances (Brown and Witter, 1979). Burning characteristics may also be different.

As indicated in Table 8.15, the hydrogen content of product gas from coal gasification will be much higher in H_2 than natural gas. Whether this higher hydrogen content will be significant in terms of hydrogen embrittlement of existing pipelines and gas-holders, or a hazard in industrial and domestic burners, remains to be determined. Table 8.15 also shows higher CO concentrations in low-Btu product gas than in natural gas. Evaluation of the potential environmental health consequences to users of these and other hydrogenation products requires more detailed information about the composition and variability of such products and the chemical and physical changes they may undergo during combustion or further chemical processing.

3 Control and Planning

Most of the problems of control in hydrogenation plants are probably solvable, providing that the nature and source of the pollution are properly described. In addition, low-sulphur refined fuels produced by such processes are much more acceptable in urban areas than raw coal or heavy, high-sulphur oils. However, because of the large quantities and complex nature of the conversion process,

TABLE 8.15 *Coal gasification product gas composition (vol. %)*

	Natural gas[a]	Wellman-Galusha[b] (Dry product gas)	Lurgi[c] (after CO conversion)	Koppers-Totzek[d] (shifted stream)	Hygas Pilot Plant[e]	Estimated Lurgi[c] (Shifted methanized stream)
		Glen Cery, York, Pa., U.S.A.	Usti, Czechoslovakia	Modderfontein, South Africa	Chicago U.S.A.	El Paso, U.S.A.
CH_4	99.2	0.2	9.9	0.1	67.4	92.9
H_2	–	16.3	39.3	54.6	15.5	4.2
CO_2	0.2	5.5	32.7	41.3	0.0	1.8
CO	–	25.5	13.4	3.0	0.0	0.01
N_2 + (Ar)	0.6	51.6	2.2	1.0	17.1	1.1
C_2H_6, C_2H_4, etc.	–	–	0.7	–	0.0	–
H_2S	–	690 ppm	1.5	–	–	–
COS	–	93 ppm	–	–	<1000 ppm	–
CS_2	–	<1 ppm	–	–	–	–
O_2	–	–	0.3	–	–	–
Lower heating value						
kcal m^{-3} at 0°C	8940	1207	–	1508	6168	8720
(Btu ft^{-3})	(953)	(136)	–	(169)	(693)	(980)

[a]Sinor (1979).
[b]Thomas, Trede and Page (1979).
[c]Mácek, Prášek and Šimánek (1976).
[d]Franzen and Goeke (1976).
[e]Ghassemi, Crawford and Quinlivan (1978).

there must be a recognition that these new types of operations may produce kinds of pollutants with which we are unfamiliar.

Available process stream data for full-scale and pilot plants, although meagre, show significant concentrations of a number of pollutants which are recognized hazards to lung and skin, and some which are known carcinogens. A major effort is needed to further characterize the wastes and products of the hydrogenation process, particularly in terms of toxic materials which may appear in low concentrations. Complete quantification of pollutant discharges requires long-term operation of large-scale pilot plants. (This recommendation is particularly important because of the way pilot plants are usually operated. Due to time and economic restrictions there is a tendency to change operating conditions frequently.)

NIOSH (1978a) have stressed the importance of medical monitoring including a pre-employment medical examination; regular check-ups (more frequent and in more detail for high-risk groups); long-term follow-up of high-risk individuals after transfer to other job categories or termination of employment; full record keeping, including work history and exposure data.

Protective procedures should include: the prompt clean-up of spilled material; prevention of leaks and regular maintenance of pollution-control equipment; proper protective clothing (Ketcham and Norton, 1960); provision of washing facilities with time allowed for their use; specific and detailed operating instruction and work practices for preventing hazardous exposures, particularly during shut-down, start-up and vessel and tank clean-out; thorough worker education on the potential hazards of the hydrogenation process; and regular and frequent monitoring of hazardous materials in work areas.

Another important concern is one of educating plant and regulatory personnel to occupational and environmental problems which are caused by release of hazardous materials from coal processing. The most effective control device available still is turned on or off by a human being whose judgement is subject to error. Chemical plant managers and operators are used to periodic upset conditions with consequent discharges to the environment. Their major concern is to return the plant to productive operation, sometimes without sufficient concern for the types and quantities of emissions. There is a need to build into both the design and operation of such plants a consciousness of minimizing or eliminating the environmental release of any stream which has a reasonable potential for containing toxic substances.

Considerable effort should be expended to review, monitor and protect the occupational health of plant workers, particularly in view of the lack of any long-term studies on occupational health effects (NIOSH, 1978b). Retrospective epidemiological investigations on identified groups of workers at pilot or commercial-sized installations where medical records exist (pre- and post-employment medical examinations) would be extremely useful in trying to better delineate long-term health implications. Continuous monitoring of pollution levels in the worker, user and community setting, as well as perspective epidemiological evaluations and the development and maintenance of pertinent disease registries on these populations, is a necessary concomitant to responsible development of coal resources.

Control of pollutants from an established technology is often difficult and costly. It is much more desirable to identify and quantify hazards and develop appropriate characterization and control methods, before a new technology is launched. Designers, operators and regulators must be aware and alert to the problems of coal hydrogenation in order to properly protect the health of the worker and the public.

4 References

ALMAULA, B. and NARDELLA, J. (1978). *Environmental Development Plan for the Coal Gasification Program FY 1977*. U.S. Department of Energy, Washington, DC.

ANTIZZO, J. (ed.) (1978). *Health and Environmental Effects of Coal Gasification and Liquefaction Technologies*. DOE/HEW/EPA–01, M78–58. The Mitre Corporation, McLean, Virginia.

BOLTON, N. E. (1976). *Industrial Hygiene Procedure: Polynuclear Aromatic Hydrocarbons*. Oak Ridge National Laboratory, Oak Ridge, Tennessee.

BOMBAUGH, K. J., CORBETT, W. E., LEE, K. W. and SEAMES, W. S. (1980). An environmentally based evaluation of the multimedia discharge from the Lurgi coal gasification system at Kosovo. Symposium on *Environmental Aspects of Fuel Conversion Technology – V*. Environmental Protection Agency, St Louis, Missouri.

BROWN, R. and WITTER, A. (1979). *Health and Environmental Effects of Coal Gasification and Liquefaction Technologies*. A Workshop Summary and Panel Reports for the Federal Inter-Agency Committee on the Health and Environmental Effects of Energy Technologies. Department of Energy Report DOE/HEW/EPA–03, MTR–79W00137, Mitre Corporation, McLean, Virginia.

CAMERON, D. S. (1974). Treating problems associated with coal gasification, El Paso Natural Gas Co., El Paso, Texas. Proceedings of *Gas Conditioning* Conference. ERDA Document No. CONF–740365.

C&EN (1979). Underground coal gasification: a crucial test. *Chemical & Engineering News*, 3 December, **57** (49), 19–30.

DOLL, R., FISHER, R. E. W., GAMMON, E. J., GUNN, W., HUGHES, G. O., TYREE, F. H. and WILSON, W. (1965). Mortality of gasworkers with special reference to cancers of the lung and bladder. *British Journal of Industrial Medicine*, **22**, 1–12.

DOLL, R., VESSEY, M. P., BEASLEY, R. W. R., BUCKLEY, A. R., FEAR, E. C., FISHER, R. E. W., GAMMON, E. J., GUNN, W., HUGHES, G. O., LEE, K. and NORMAN-SMITH, B. (1972). Mortality of gasworkers: final report of a prospective study. *British Journal of Industrial Medicine,* **29**, 394–406.

FRANZEN, J. E., and GOEKE, E. K. (1976). Gasify coal for petro-chemicals. *Hydrocarbon Processing*, **55**, (11), 134–8.

GHASSEMI, M., CRAWFORD, K. and QUINLIVAN, S. (1978). *Environmental Assessment Data Base for High-Btu Gasification Technology:* Vol. II, Appendices, A, B and C. U.S. Environmental Protection Agency Report EPA–600/7–78–186b, Washington, DC.

GHASSEMI, M., CRAWFORD, K AND QUINLIVAN, S. (1979). *Environmental Assessment Report: Lurgi Coal Gassification Systems for SNG.* U.S. Environmental Protection Agency Report EPA-600/7-79-120, Washington, DC.

GLUSKOTER, H. J., RUCH, R. R., MILLER, W. G., CAHILL, R. A., DREHER, G. B. and KHUN, J. K. (1977). *Trace Elements in Coal: Occurrence and Distribution.* ISGS Circ. no. 499, Illinois State Geological Survey, Urbana, Illinois.

HEMENWAY, A. and ALMAULA, A. (1978). *Environmental Development Plan for the Coal Liquefaction Program FY 1977.* U.S. Department of Energy, Washington, DC.

HOLMES, J. G., KIPLING, M. D. and WATERHOUSE, J. A. H., (1970). Subsequent malignancies in men with scrotal epithelioma. *Lancet,* 25 July, pp. 214–15.

KAWAI, M., AMAMOTO, H. and HARADA, K. (1967). Epidemiological study of lung cancer. *Archives of Environmental Health,* **14**, 854–64.

KETCHAM, N. H. and NORTON, R. W. (1960). The hazards to health in the hydrogenation of coal. III. The industrial hygiene studies. *Archives of Environmental Health,* **1**, 194–207.

LAWTHER, P. J., COMMINS, B. T. and WALLER, R. E. (1965). A study of the concentrations of polycyclic aromatic hydrocarbons in gaswork retort houses. *British Journal of Industrial Medicine*, **22**, 13–20.

LLOYD, J. W. (1971). Long-term mortality study of steelworkers. V. Respiratory cancer in coke plant workers. *Journal of Occupational Medicine*, **13**, 53–67.

MÁCEK, J., PRÁŠEK, K. and ŠIMÁNEK, K. (1976). Ecological aspects of gas production from high-sulphur lignite. Paper IGU/87–76, 13th World Gas Conference, London.

MAZUMDAR, S. C., REDMOND, C., SOLLECITO, W. and SUSSMAN, N. (1975). An epidemiological study of exposure to coal tar pitch volatiles among coke oven workers. *Journal of the Air Pollution Control Association*, **25**, 372–9.

MORRIS, S. C., MOSKOWITZ, P. D., SEVIAN, W. A., SILBERSTEIN, S. and HAMILTON, L. D. (1979). Coal conversion technologies: some health and environmental effects. *Science*, **206**, 654–62.

NAS (1972). *Particulate Polycyclic Organic Matter.* National Academy of Science, Washington, DC.

NIOSH (1976a). *Criteria for a Recommended Standard: Occupational Exposure to Benzene*. National Institute for Occupational Safety and Health, Washington, DC.

NIOSH (1976b). *Criteria for a Recommended Standard: Occupational Exposure to Carbon Dioxide*. National Institute for Occupational Safety and Health Report 76–194, Washington, DC.

NIOSH (1978a). *Recommended Health and Safety Guidelines for Coal Gasification Pilot Plants.* National Institute for Occupational Safety and Health Publication 78–120, Cincinnati, Ohio.

NIOSH (1978b). *Criteria for a Recommended Standard: Occupational Exposures in*

Coal Gazification Plants. National Institute for Occupational Safety and Health Publication 78–191, Cincinnati, Ohio.

PALMER, A. (1979). Mortality experience of 50 workers with occupational exposures to the products of coal hydrogenation processes. *Journal of Occupational Medicine*, **21**, 41–4.

PARKER, C. L. and DYKSTRA, D. I. (1978). *Environmental Assessment Data Base for Coal Liquefaction Technology*, vol II. Synthoil, H-Coal, Exxon, Donor Solvent Processes. U.S. Environmental Protection Agency Report EPA–600/7–78–184b, Washington, DC.

PARTRIDGE, L. J. (1976). Coal-based ammonia plant operation. *Chemical Engineering Progress*, **56**, (8), 57–61.

PATTERSON, R. K. (1980). Ambient air downwind of the Kosovo gasification complex: *A Compendium Symposium on Environmental Aspects of Fuel Conversion Technology – V.* U.S. Environmental Protection Agency, St Louis, Missouri.

PRÁŠEK, K. (1980). *Effects of Coal Conversion by Gasification on the Environment.* Fuel Research Institute, Bechovice.

REDMOND, C. K., CIOCCI, A., LLOYD, J. W. and RUSH, H. W. (1972). Long-term mortality of steelworkers. VI. Mortality from malignant neoplasms among coke oven workers. *Journal of Occupational Medicine*, **14**, 621–9.

SCHILLER, J. E. (1977). Composition of coal liquefaction products. *Hydrocarbon Processing*, **65** (1), 147–52.

SCHLINGER, W. G. and RICHTER, G. N. (1980). Process pollutes very little. *Hydrocarbon Processing*, **59** (10), 66–70.

SCHULTZ, W. D. (ed.) (1976). *Preliminary Results: Chemical and Biological Examination of Coal Derived Materials.* Oak Ridge National Laboratories Report ORNL/NSF/EATC–18. Oak Ridge, Tennessee.

SEXTON, R. J. (1960a). The hazards to health in the hydrogenation of coal. I. An introductory statement on general information, process description, and a definition of the problem. *Archives of Environmental Health*, **1**, 181–6.

SEXTON, R. J. (1960b). The hazards to health in the hydrogenation of coal. IV. The control programme and the clinical effects. *Archives of Environmental Health*, **1**, 208–31.

SHABAD, L. M. and DIKUN, P. P. (1959). *Air Pollution by Carcinogenic Substances –3,4,–Benzpyrene.* Medghiz, Leningrad.

SHABAD, L. M., KHESINA, A. Y. and KHITROVO, S. S. (1968). On carcinogenic hydrocarbons in the exhaust gases of automobile engines and the possibility of their removal. *Vestnik Akademii Meditsinski Nauk U.S.S.R.*, **23** (6), 12–20.

SINOR, J. E. (1979). Reconsider coal conversion processing. *Hydrocarbon Processing*, **58** (7), 159–62.

THODE, H. C. JR. and CROWTHER, M. A. (1980). *Solid Waste Module for Coal Gasification.* Draft BNL Report.

THOMAS, W. C., TREDE, K. N. and PAGE, G. C. (1979). *Environmental Assessment: Source Test and Evaluation Report-Wellman-Galusha (Glen Gery) Low Btu Gasification.* U.S. Environmental Protection Agency Report EPA–600/7–79–185, Washington, DC.

WADDEN, R. A. (ed.) (1978). *Energy Utilization and Environmental Health: Methods for Prediction and Evaluation of Impact on Human Health.* Wiley, New York.

WEIL, C. S. and CONDRA, N. I. (1960). The hazards to health in the hydrogenation of coal. II. Carcinogenic effect of materials on the skin of mice. *Archives of Environmental Health*, **1**, 187–93.

9

Impacts of Coal on Natural Environmental Systems

1 Introduction

Air pollution disasters such as those in the Meuse Valley in Belgium, Donora, Pennsylvania and the city of London in the 1930s, 1940s and 1950s led to considerable research and development activity in atmospheric pollution control. The use of cleaner fuels and/or higher combustion temperatures, the installation of better emission control equipment, the construction of taller chimneys, the use of an intermittent episode control strategy and sensible land-use planning all combined to engender considerable optimism that serious air pollution would eventually be a thing of the past.

This faith in technology is not so prevalent today. There is now a growing awareness that new problems exist that are less amenable to technological solutions and that changes in the balance of fuel utilization will only add to these. The major concerns are as follows:

Photochemical air pollution

It was found by Haagen-Smit in the 1950s that several trace substances (e.g., NO_x and hydrocarbons) react photochemically to create harmful secondary pollutants (e.g., oxidants). Important examples are to be found in the Los Angeles-type smog, in possible depletion of the stratospheric ozone layer by chlorofluorocarbons, and in the conversion of SO_2 and NO_x gases to sulphate and nitrate particles (Renzetti and Doyle, 1960; Sidebottom *et al.*, 1972). MacFarland (1975) has reviewed this type of air pollution.

Carbon dioxide greenhouse warming

CO_2 has never been classified as an air pollutant, and has therefore not been the subject of intensive control technology research and development. Yet present trends in the conventional use of fossil fuels lead climatologists to expect at least a doubling of atmospheric CO_2 concentrations by the middle of the next century and in addition a recent estimate of CO_2 emissions from synthetic fuel supply cycles are put at 40% more than for direct coal combustion alternatives (Woodwell *et al.*, 1979). The implications for climate change are examined in Section 3.

Acid rain

The Swedish Case Study for the 1972 UN Conference on the Human Environment (Engström, 1971) showed that the atmosphere was a "carrier" of sulphates and nitrates from urban and industrial regions to pristine lakes and rivers sometimes 500–1000 km distant. The substances lower the pH of precipitation and this has a long-term effect on sensitive water bodies (i.e. lakes and rivers with low buffering capacity). This type of water is found commonly in Scandinavia and in the Laurentian Shield of North America, where there has been a gradual decline in the pH of affected lakes and rivers. These questions are examined in Section 4.

Increasing coal utilization

With the rising price of oil, there will be increasing use of coal in coming decades. This is likely to have three major environmental effects:

(1) For power stations which convert from oil to coal, existing environmental control programmes may no longer be adequate. As a result of changes in stack gas composition, exit velocity and gas temperature, the pattern of downwind ground-level concentrations may change within a few tens of kilometres from the site.
(2) For new power stations, the emissions of SO_2, NO_x and CO_2 will contribute to the problems of acid rain and "greenhouse" climate warming. The emissions are expected to be slightly larger per unit of energy generated from an oil-fired or gas fired station.
(3) For the newer coal conversion technologies (gasification and liquefaction) a new "mix" of environmental problems will arise for which there needs to be new control expertise on a wide scale.

Industrial activities other than coal-fired power generation contribute to the problems listed above, and it is not possible to discuss the impact of coal utilization in isolation but it is possible to describe in general terms the impacts of CO_2, SO_2 and NO_x on local, regional and global environments. It is then possible to attempt an estimate of the incremental harm likely to result from increased coal utilization in the next several decades. In order to do this Section 2 presents general discussion of past, present and future emissions of fossil fuels, with special reference to the contribution of coal to these emissions.

2 Emissions from the Combustion of Fossil Fuels

Estimates of future CO_2, SO_2 and NO_x emission trends are necessarily speculative since it is impossible to predict future coal, oil and gas consumption with any accuracy. Factors such as environmental and social constraints, the success or otherwise of recovering oil from oil sands and shales, the finding rate of crude oil and the development of nuclear power will all influence the supply of, and demand for, coal.

2.1 Past Trends

Keeling (1973) has compiled an estimate of world production of fossil fuels until 1969. This has been updated to 1976 by Rotty (1979) who estimated that in 1976 there were 5038×10^6 t of carbon emitted as CO_2, of which approximately 40% was derived from coal. The upper part of Table 9.1 shows the upward trends between 1860 and 1976. Coal entirely dominated the CO_2 emissions until the 1940–50 period when substantial oil and natural gas production resulted in a slowing of the coal production rate. By 1969 coal emissions, although larger than at any time previously, had been overtaken by petroleum and represented some 43% of the total.

2.2 Future CO_2 Emissions

Table 9.1 includes speculative estimates of future emissions of CO_2. It is possible that the fraction of CO_2 from coal will "bottom out" at about the present level (approximately 37%), then increase steadily at 5–10% per year. These growth estimates are somewhat lower than those of Rotty (1979). Discovery of new large petroleum basins could completely alter the situation.

TABLE 9.1 *Estimates of past and future carbon emissions as CO_2 in units of 10^6 t*

Year	Coal and lignite	Oil		Gas	Total	Percentage from coal
1860	93	0.1		0	93	100
1880	224	3		0	225	99
1900	505	16		4	525	96
1920	870	76		13	959	91
1929	982	159		30	1171	84
1939	974	220		40	1234	79
1949	1022	358		89	1469	70
1959	1445				2416	60
1969	1598				3688	43
1976	1936		3102		5038	38
1980	2070		3500		5570	37
1990	2850		3800		6650	42
2000	3930		4000		7930	50
2010	5410		3700		9110	59
2020	7445		3400		10 845	69
2030	10 250		3100		13 350	77

Data for 1960–69 from Keeling (1973).
1980–2030 data are speculative and assume 3.2% annual growth for coal.

2.3 SO_2 and NO_x Emissions

Robinson and Robbins (1968) have estimated that 146 × 10^6 t of SO_2 are emitted annually by anthropogenic activities. Of this 102 × 10^6 t (70%) are from coal burning, 28.5 × 10^6 (16%) from petroleum and the remaining 14% from smelting of sulphide ores. Biological decay processes on land and in water generate substantial quantities of H_2S, possibly amounting in total to three times the anthropogenic sulphur emissions (Eriksson, 1963; Robinson and Robbins, 1968) but these natural emissions are widely distributed whereas the anthropogenic emissions are concentrated in industrialized areas, notably in North America and Europe.

From estimates of fossil fuel production for 1965 (Keeling, 1973) and the WHO (1979) quoted SO_2 emissions for that year, it is possible to calculate an emission rate of SO_2 from solid and petroleum fuels. This amounts to 102 × 10^6 t SO_2 (51 × 10^6 t S) from

2268×10^6 t coal or 0.022 t S per ton coal. This emission factor depends significantly on the sulphur content of the coal, on the extent of retention of sulphur in the ash and on SO_2 removal practices. There is little doubt that this factor could be reduced substantially. U.S. performance standards of 1.2 lb SO_2 per 10^6 Btu have been promulgated for 1980 which at 12 000 Btu lb^{-1} coal amounts to 0.014 t SO_2 or 0.007 t S per ton coal. The corresponding emission factor for NO_x is one-third that of SO_x (Chem. Eng., 1977).

The 1965 emissions from oil and gas were 28.5×10^6 t from 1539 $\times 10^6$ t oil and 704×10^9 m^3 gas. Keeling (1973) quotes 0.84 t C per t of carbon emitted. This quantity could undoubtedly be reduced 380×10^6 t from gas, a total of 1672×10^6 t. An approximate emission factor for SO_2 petroleum is thus 0.017 t SO_2 (0.0085 t) per ton of carbon emitted. This quantity could undoubtedly be reduced by sulphur removal technologies. Most sulphur in oil is present as organic sulphur in the heavy fuel oil fractions which are used for heating by direct combustion. The sulphur in natural gas is mainly H_2S which is easily separated by the Claus process.

It is impossible to predict SO_2 and NO_x emissions accurately but it seems likely that they will "track" the CO_2 levels, increasing to approximately double their present values by the year 2000 and doubling again by 2025. Widespread use of sulphur removal technology applied to coal as fuel or to the stack gases could reduce these levels but this may be offset by the tendency to exploit low-sulphur coals first, leaving the high-sulphur coals until later. Perhaps the most significant determinant will be the price which societies will wish to pay for SO_2 and NO_x emission reduction. The price will be a function of the perceived adverse environmental and health effects juxtaposed with the adverse affects of energy costs and shortages.

It can be concluded that without the introduction of sulphur removal technology, SO_2 emissions will at least increase in proportion to the overall increase in fossil fuel consumption. There is a potential for these emissions to increase faster due to the higher proportion of coal used and the early exploitation of low-sulphur coals. An increase in SO_2 emissions by a factor of 5 is thus possible by 2025. NO_x emissions will also tend to increase in proportion to fossil fuel consumption. By making a very significant and expensive technical effort, it may be possible to reduce the emissions of SO_2 and NO_x per unit of fuel burned to a level such that the total tonnage of emissions of these compounds per year remains approximately constant.

2.4 *Suspended Particulates*

On a global scale, and even on a regional scale, the emissions of particulate matter from power stations are small in comparison with other man-induced and natural emissions, although acid sulphate aerosols may make a significant contribution to the overall particulate load.

2.5 *Trace Elements*

It will be appreciated, from the account of the environmental significance of trace elements from coal combustion and conversion processes (Chapter 6) that it is not possible to make accurate predictions of local, regional and global emissions of trace elements from increased coal utilization. However, increases will occur and locally, at least, these will give rise to increased levels of certain elements in food chains (Novakova, 1969; Novakova and Paukert 1980).

3 The Climate Problem

3.1 *The Elements of the CO_2 Problem*

There is no doubt that atmospheric CO_2 concentrations are rising. Figure 9.1 shows the trend at Mauna Loa, Hawaii (Keeling and Bacastow, 1977). In addition to a well-defined annual cycle (due to the uptake of CO_2 by vegetation during the growing seasons and re-release into the atmosphere when the vegetation is dormant), there is an unmistakable upward trend of about 1 ppm yr^{-1}. Other global CO_2 monitoring stations exhibit similar behaviour.

A continuing rise in atmospheric CO_2 concentrations is certain. There remain only the questions of the extent of the increases in future decades and the resulting impacts on climate and on society. More specifically, the following components need to be examined (Munn and Machta, 1979):

(1) rate of introduction of CO_2 into the atmosphere from fossil fuel combustion (discussed in Section 2 above);
(2) rates of deforestation, reafforestation and cultivation of virgin lands;
(3) modelling of the global biogeochemical cycle of carbon;

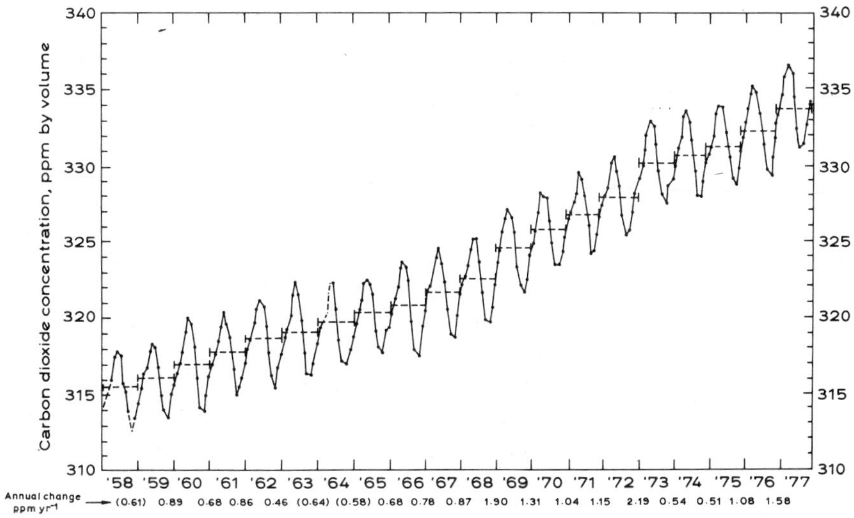

FIGURE 9.1 *The mean monthly atmospheric CO_2 concentrations at Mauna Loa, Hawaii (Keeling and Bacastow, 1977). Dashed horizontal lines are annual concentrations.*

(4) prediction of atmospheric levels of CO_2 for various scenarios of future input of CO_2;
(5) prediction of climatic changes due to increased atmospheric CO_2;
(6) examination of other possible causes (natural and man-induced) of concurrent climate change and estimation of the net effects;
(7) analysis of the impact of climatic change and enhanced CO_2 on the biosphere and on human activities;
(8) engineering and socioeconomic studies of technological fixes.

3.2 *The Land Biosphere*

In terrestrial ecosystems, particularly the world forests, there has been an approximate balance over centuries between the amount of carbon assimilated from the atmosphere by photosynthesis and the amount returned by respiration and fire. The factors that have changed recently are:

(1) the rates of deforestation, reafforestation and cultivation of virgin lands (the first and third of these processes are sources of atmospheric CO_2, the second is a sink);

(2) the current rise in atmospheric CO_2 concentrations; in principle, this "fertilizes" the world biosphere and thus creates a small net sink for CO_2.

In a recent overview of this question, Emmanuel, Killough and Olson (1980) are of the opinion that the net flux of CO_2 from the global biosphere to the atmosphere was near zero in 1970, with approximately 113 Gt yr^{-1} flowing in each direction. Emmanuel, Killough and Olson (1980) believe that the effect of factor (1) was insignificant; with respect to factor (2), they are of the opinion that the terrestrial carbon system is so efficient in transferring photosynthesized carbon to longer-storage compartments of the biosphere that the effect of increased photosynthesis on atmospheric concentrations, at least between 1860 and 1970, was negligible.

Recent rates of deforestation, reafforestation and cultivation of virgin lands are not well known but presumably they should result in a new small net source of atmospheric CO_2. They should also produce important regional differences; for example, deforestation is taking place faster in the tropics than in the temperate zones. But even in the absence of these disturbances, scientists are beginning to think that temperate forests may act as a general sink for CO_2, to be balanced by an equivalent source in the tropical rain forests (DOE, 1980).

3.3 The Global Biogeochemical Cycle of Carbon

Averaged over the globe and over time-periods of a year or longer, the increase in atmospheric CO_2 shown in Figure 9.1 amounts to about one-half of the amount of CO_2 added to the atmosphere by fossil fuel combustion. If the land biosphere is neither a net source nor a net sink for CO_2, then the remaining 50% is presumably taken up by the world oceans. However, in order to understand the processes concerned and to make predictions, it is necessary to model the global biogeochemical cycle of carbon, including its pathways through the terrestrial, freshwater and marine components. The usual approach is to imagine that the world consists of a series of connecting boxes (e.g. land biosphere, lower atmosphere, etc.), each pair having a characteristic exchange coefficient to be determined experimentally. The transfer rates between boxes are poorly known but the models all suggest that the main bottleneck to the removal of excess CO_2 from the atmosphere is transfer from marine surface waters to the deep ocean.

The real world is not as simple as present models suggest. For example, there is recent experimental evidence (DOE, 1980) that the equatorial Atlantic, Pacific and Indian Oceans are sources rather than sinks of CO_2. Sea ice too is believed to be a small net source. Present efforts are therefore directed towards developing three-dimensional models of the global carbon cycle. Some recent publications review this topic (Williams, 1979; SCOPE, 1979; Bolin, 1980).

3.4 *Predicting Future Atmospheric CO_2 Concentrations*

In order to predict future CO_2 concentrations, an assumption must be made about the rate of consumption of fossil fuels in future decades. This is subject to uncertainties. Nevertheless, for various biogeochemical models of carbon and for various scenarios of fossil-fuel use, there is a consensus amongst the modellers that in the most general terms, pre-industrial CO_2 concentrations may well double by the early to mid-part of the next century (see, for example, Keeling and Bacastow, 1977). Viewed in a long-term perspective, the model used and the energy consumption assumed make little difference in the timing and magnitude of the peak concentration. This is the reason for current concern by atmospheric scientists, who are urging that this question be given high research priority, and that long-term options should be kept open by the power-generation sector with respect to controlling CO_2 emissions.

3.5 *Predicting the Resulting Effects on Climate*

The climatic impact of increasing atmospheric CO_2 concentrations has traditionally been called "greenhouse warming". Downward-streaming solar radiation is not attenuated by CO_2 whereas upward-flowing terrestrial infra-red radiation is absorbed in selected wave-bands of the CO_2 spectrum. If all other factors were to remain unchanged, the net effect of increasing CO_2 would therefore be to warm the lower atmosphere. Most atmospheric climate models predict that if CO_2 concentrations were to double and feed-back mechanisms were ignored, the mean temperature of the lower atmosphere would rise by about 1.5–3°C (see Munn and Machta (1979) for a summary). These changes would not be seasonally or geographically uniform (Manabe and Stouffer, 1979; Manabe and Wetherald, 1980) but there is no doubt that this "would be a significant alteration of the present climate, unprecedented in historical times" (DOE, 1980). Not only would the temperature change, but also wind, cloud and precipitation patterns would be affected. In some sub-polar regions surface temperatures might actually cool and glaciers might build as a result of increased moisture transfers from an ice-free ocean.

Two additional points should be mentioned. In the first place, it should be emphasized that although CO_2 concentrations have been increasing over the last century, there is as yet no evidence that climate has been affected. This may be due to the natural variability of climate which makes it difficult to detect trends. In fact, one of the major problems associated with CO_2 warming is the uncertainty in distinguishing a naturally occurring long-term fluctuation in temperature from an irreversible upward trend.

Secondly, two recent predictions (Newell and Dopplick, 1979; Idso, 1980) suggest that a doubling of CO_2 concentrations would cause world temperatures to rise by less than 0.25°C. This result is not accepted by most climate modellers and the Idso (1980) prediction has been the subject of a rebuttal.

3.6 *Other Possible Causes of Climatic Change*

Gases which contribute to greenhouse warming (although to a lesser extent than CO_2) include chlorofluorocarbons, N_2O, ammonia and methane (see Munn and Machta, 1979). Atmospheric concentrations of these substances are increasing, which should result in an amplification of greenhouse warming.

A second major factor suggested as a cause of climatic change is suspended particulate matter, which is lifted into the atmosphere by soil erosion, sea spray, forest fires, volcanoes, agricultural burning and industrial processes. There have been regional changes in particulate concentrations in recent years but it is inappropriate to discuss global trends in the lower atmosphere because particle residence times are too short (5 days on average) to permit global dispersion. Downwind of deserts and deforested areas, the populations of quartz and soil particles may have increased in some cases. In addition, it can be stated with certainty that downwind of industrial areas, haziness has been increasing in the last 30 years (Munn, 1973); the haze is believed to be composed mainly of sulphate and nitrate particles, formed photochemically from SO_2 and NO_x gases, 500–1000 km downwind of source regions.

It has been postulated that particulates (mainly <5 μm in diameter) in the lower atmosphere cause cooling, but this question is far from settled. The present majority view is that over the continents, particles actually warm the lower atmosphere; over the oceans, Munn and Machta (1979) suggest that particles may either warm or cool their surroundings depending on whether they are reflecting (e.g. quartz) or absorbing (e.g. carbon).

Volcanic eruptions of particulate matter and SO_2 (which is later converted to sulphates) are believed to be of greater importance than are tropospheric particles because of the long time the stratosphere takes to purify itself. Bryson and Goodman (1980), for example, are convinced that cool decades are associated with increased global volcanic activity.

Finally, there are the very long-term climatic oscillations due to astronomical factors. Imbrie and Imbrie (1980) have suggested, on the basis of the Milankovitch theory, that the earth is undergoing a long-term cooling trend which began 6000 years ago and which will continue for another 23 000 years. If this is so, then CO_2 warming may be welcomed!

3.7 The Impact of Greenhouse Warming on the Biosphere and on Society

There is laboratory evidence that when other environmental stresses are not limiting, photosynthesis and hence agricultural and forest productivity will increase with increasing atmospheric CO_2 concentrations. In addition, a rise in temperature (provided that it is not accompanied by a decrease in precipitation) will push poleward the frontiers for agriculture and forestry.

Predicting the effect of climatic change on society is a much more subtle problem. Whereas the biosphere is generally able to absorb environmental stresses without great difficulty, society is rather fragile, almost totally dependent on technology in the industrialized countries and almost totally dependent on the vagaries of climate in the Third World. Technology optimizes life-support systems such as agriculture over the short term and often cannot cope with large swings in climate, because of underlying socioeconomic practices and structures. It should also be mentioned that society's perception of environmental change is often more important than the actual change. For example, a government would respond differently if it believed that a drought was a temporary anomaly than it would if it thought that there had been a permanent shift to drier conditions.

Climate change *per se* is not particularly bad. However, there must be concern about society's ability to cope with climate change, or perceived changed. Studies of the socioeconomics of climate change are only beginning, and should be accelerated. This challenge has been recognized by the World Meteorological Organization and UNEP through the World Climate Programme.

3.8 Technological Fixes

If it were felt essential, several technological fixes could be considered to slow down the rise in CO_2 (Munn and Machta, 1979; Steinberg and Albanese, 1980):

(1) Substitute alternate energy sources for fossil fuels. (The time required for such a transition is many decades; furthermore, the alternatives may generate undesirable environmental impacts.)
(2) Removal of CO_2 from major sources. (This is feasible but large quantities of energy would be required. For 50% CO_2 removal, power plant efficiency decreases from 34 to 25%; for 90% removal, efficiency drops to between 15 and 6% (Steinberg and Albanese, 1980).)
(3) Removal of CO_2 from the atmosphere and/or the oceans by chemical and cryogenic methods. (The CO_2 must be stored. The energy demands for removal may exceed the amount of energy derived from the burning of fossil fuels.)
(4) Increase the biosphere by planting more trees. (For projected CO_2 growth, the number of trees to be planted is prohibitive; when cut, the wood must be sequestered in "permanent" non-exchangeable reservoirs.)
(5) Accelerate oceanic uptake by dumping carbonates into the deep ocean. (The costs would be substantial.)
(6) Counteract CO_2-induced climatic changes by another form of human intervention. (The addition of dust or aerosols over the oceans has been suggested, for example.)

3.9 The Impact of Increasing Coal Utilization on the CO_2 Problem

Table 9.1 indicated that the contribution of coal to fossil fuel emissions is expected to increase in the next 50 years, rising to more than 70% before the middle of the next century. In absolute terms too, coal consumption is expected to increase substantially, causing CO_2 emissions to rise whereas little change is anticipated in emissions from oil and gas. Despite great uncertainty in the numerical values of these projections of increased coal utilization, there is no reason to doubt the general view that the expected rapid rise in atmospheric CO_2 concentrations in the next 50 years will be largely due to increasing use of coal.

4 SO_2/NO_x Effects on the Biosphere

4.1 The Dispersion of Chimney Plumes

Several factors affect the dispersion of pollutants released into the atmosphere from tall chimneys.

Initial conditions

A chimney plume usually rises at first, because it is hot and has an exit velocity. After a few kilometres, the plume becomes horizontal.

Transport

The wind carries the plume with it, sometimes more than 1000 km.

Turbulent diffusion

Wind gusts mix the plume in the crosswind and vertical directions.

Chemical transformations

In sunlight, SO_2 and NO_x are gradually converted to sulphate and nitrate particles (at a rate of from 1 to 5% hr^{-1}).

Dry deposition

Large particles are quickly removed from the atmosphere by gravity fall-out. Small particles and gases are slowly removed by adsorption on to vegetation, buildings, lakes and other surfaces.

Wet deposition

Through capture by cloud droplets, rain or snow, pollutants are scavenged from the atmosphere. Precipitation becomes acidic when it takes up sulphates and nitrates, releasing hydrogen ions in solution.

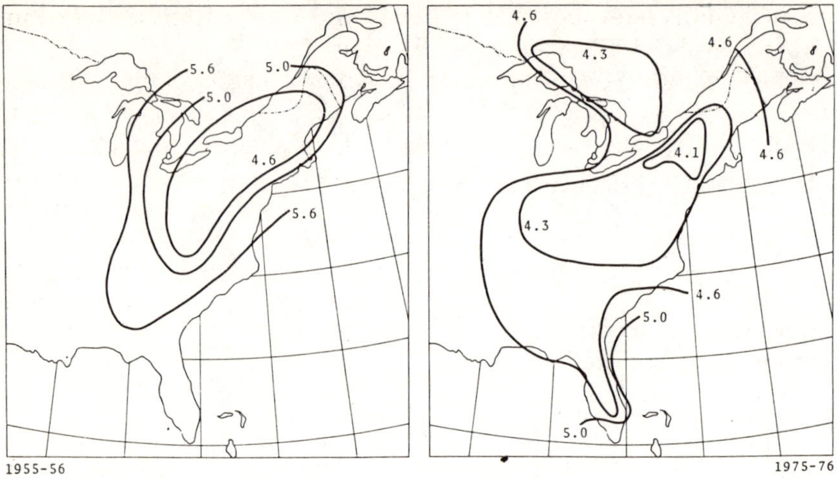

FIGURE 9.2 *Isopleths showing annual average pH for precipitation in eastern North America (modified from Likens et al., 1979).*

Over long distances and averaged over a year or longer, the dry and wet deposition processes contribute about equally to the removal of chimney plumes from the atmosphere.

Because of increasing emissions of SO_2 and NO_x over the last several decades, there has been a decrease in the pH of precipitation downwind of large industrial regions. Figure 9.2 shows the changes that have occured in eastern North America since the mid-1950s (Likens *et al.*, 1979), and the situation is similar in Scandinavia (Wallen, 1980). In the North American case, two-thirds of the acidity is due to SO_2 and one third to NO_x (U.S.A./Canada, 1979).

In order to predict future trends and to estimate the relative contribution of coal-fired power stations, the usual approach is to develop an atmospheric long-range transport model. Current models generally contain the following components: emission source inventories; turbulent transport and diffusion; chemical transformations; and wet and dry deposition. Once the model has been validated, the relative impacts of various source types can be determined. In addition, the impacts of alternative scenarios and control strategies can be examined. The main difficulty with this approach is that the atmosphere is so variable, from day to day and from year to year, that there is considerable difficulty in verification and making projections, except in some average sense.

Current models of long-range transport include sulphur chemistry, but are deficient in their treatment of the oxides of nitrogen. Rodhe, Crutzen and Vanderpol (1979) suggest that there are important interactions between sulphur and nitrogen photochemistry, which means that one may not assume a proportional dependence of sulphates and nitrates on emission rates of SO_2 and NO_x. It seems that a strategy aimed at controlling sulphur alone should be viewed with suspicion at this stage of our knowledge of atmospheric chemistry.

4.2 *Direct Effects of SO_2 and NO_x* on Vegetation

Elevated concentrations of SO_2 have long been recognized as potentially phytotoxic, and the plant damage caused by SO_2 has recently been summarized (Jeffree, 1979). The inability of certain species to grow in urban and industrial centres has been attributed to this phytotoxicity. The absence of lichens, for example, in major cities relates to their sensitivity to atmospheric SO_2 concentration (Gilbert, 1965; LeBlanc and DeSloover, 1970).

In the past, concern has focused on relatively high SO_2 concentrations occurring immediately adjacent to the SO_2 source, usually an industrial plant where controls were absent or inadequate. The impact on the environment from this type of emission can be enormous. Entire ecosystems have been altered as a result of elevated SO_2 concentrations in the atmosphere. For example, 830 ha of deciduous forest were completely denuded in Copper Basin, Tennessee, resulting in erosion and microclimatic changes; an additional 6880 ha were replaced by grassland (Treshow, 1970). Similar effects have been observed at smelter locations near Sudbury, Ontario; Trail, British Columbia; Anaconda, Montana; as well as Douglas, Arizona (Katz, 1939; Treshow, 1970; Linzon, 1972; Hutchinson and Whitby, 1974).

Today taller chimneys have improved conditions locally, by committing the dose to a wider area. Acute effects on a local scale may thus be replaced by chronic effects on a regional scale.

4.2.1 Mode of toxicity

Sulphur dioxide enters plants through stomata on the leaf surface. In the stomatal cavity it is rapidly converted on the moist cell surfaces into sulphite (SO_3^{2-}) and more slowly into sulphate (SO_4^{2-}). Sulphite is a highly toxic compound, thirty times more toxic than SO_4, but high concentrations of the latter can also be toxic. The damaging effects of SO_3 will occur if SO_2 entry into the leaf exceeds the rate of conversion of SO_3 into SO_4, so that SO_3 accumulates.

Two different conditions produce toxic symptoms. The first involves high SO_2 concentrations for brief exposures. This is classified as an acute effect and occurs as a result of SO_3 accumulation. Symptoms include:

(1) rapid loss of green pigmentation (chlorosis) due to chlorophyll degradation (Rao and LeBlanc, 1966; Wellburn, Majernik and Wellburn, 1972; Sandström and Hällgren, 1973) which usually occurs along the margins or tips of leaves where SO_3 accumulates;
(2) necrosis, which results from the death of the cells and occurs primarily between the veins; in conifers, necrosis is usually present as tip burns, reddish-brown banding or as basal burns; the younger needles are generally more susceptible due to their higher metabolic rates;
(3) premature senescence or defoliation; in Douglas fir (*Pseudotsuga menziesii*) this may occur several days or weeks following injurious exposures, while in pines (*Pinus* spp.) the needles may fall after 1–3 years instead of the usual 5 years (Treshow, 1970);
(4) as a result of chlorosis, necrosis and premature senescence, there can be both reduced photosynthetic activity and reduced yield.

In temperate climates early autumn colours may develop as a result of SO_2 fumigations. Acute toxicity occurs at ambient concentrations above 0.5 ppm SO_2, although this is highly dependent upon environmental factors and is species-specific.

Exposure to low SO_2 concentrations (less than 0.1–0.3 ppm SO_2) for long periods of time can also be harmful to plants. This produces chronic effects and results from SO_3 or SO_4 accumulation. Chronic symptoms are generally not as severe as acute symptoms and develop more slowly. Symptoms include: chlorosis, reduced photosynthesis, and reduced growth, resulting in lower yields. Necrosis seldom occurs at these low SO_2 concentrations except in the case of very sensitive species.

Recently there has been some debate concerning the effects of "invisible" injury. This refers to impaired physiological functions as a result of SO_2 in the absence of any visible external symptoms. The debate has not as yet been resolved although evidence appears to suggest that even in the absence of external symptoms, rate of photosynthesis and growth can be reduced (Bell, Rutter and Relton, 1979).

Nitrogen oxides in the form of nitric oxide (NO), nitrogen dioxide (NO_2) and nitrogen tetraoxide (N_2O_4), can also be harmful to plants. Gradual exposure of plants to low NO_x concentrations (30–50 ppm) results in characteristic downward cupping of leaves and a dark green coloration due to an increase in chlorophyll content. Even at these low concentrations, plant weight and leaf size can be significantly reduced within 1 week of exposure (Treshow, 1970). Exposure to higher concentrations can produce symptoms similar to SO_2 injury: chlorosis, necrosis, and defoliation. Injury can occur within 3 hours at 4 ppm for alfalfa (*Medicago sativa*), 3 hours at 6 ppm for oats (*Avena sativa*) and 8 hours at 2 ppm for tobacco (*Nicotiana tabacum*).

4.2.2 Species sensitivity

Species vary greatly in their sensitivity to SO_2. As previously mentioned, lichens are extremely vulnerable to SO_2 pollution. One reason for this is that they lack both a cuticle and stomata. As a result, exchange of gases occurs across the entire surface. Another reason is that, unlike most plants, lichens are active during the winter. Therefore, exposure of metabolically active plants to SO_2 is not limited to the summer months (Sandström and Hällgren, 1973). Due to their SO_2 sensitivity, some lichen species can be used as indicator species for air pollution studies. Experiments involving lichen "transplants" appear to be useful in this respect. Elaborate mapping techniques in major urban areas have been correlated with SO_2 levels (Gilbert, 1965; LeBlanc and DeSloover, 1970).

Conifers are generally considered to be sensitive to SO_2, although there is considerable variation in sensitivity among this group. Eastern white pine (*Pinus strobus*), for instance, is known to be very sensitive to SO_2. Linzon (1972) studied the effect of three smelters at Sudbury, Ontario on the distribution and growth of eastern white pine. He found that within 40 km NE of Sudbury and within 10 km NNE of Falconbridge, eastern white pine was either completely lacking or severely damaged by SO_2. An estimated 1870 km^2 area showed severe foliar injury of eastern white pine.

In contrast to coniferous trees, deciduous trees are only harmed by

TABLE 9.2 *Minimal average SO_2 concentrations producing injury to various crop and other plant species (Dreisinger, 1964)*

		Maximum Average Concentration (ppm)			
		1 hr	2 hr	4 hr	8 hr
Tree species					
Quaking Aspen	*(Populus tremuloides)*	0.42	0.39	0.26	0.13
Jack Pine	*(Pinus banksiana)*	0.52	0.44	0.29	0.20
White birch	*(Betula populifolia)*	0.46	0.38	0.28	0.21
White pine	*(Pinus strobus)*	0.45	0.35	0.25	0.21
Larch	*(Larix decidua)*	0.41	0.38	0.34	0.26
Large-toothed aspen	*(Populus grandidentata)*	0.66	0.43	0.37	0.20
Willow	*(Salix spp.)*	0.41	0.38	0.33	0.30
Alder	*(Alnus spp.)*	0.46	0.43	0.43	0.21
Red pine	*(Pinus resinosa)*	0.78	0.69	0.44	0.30
Austrian pine	*(Pinus nigra)*	0.66	0.45	0.44	0.33
Hazel	*(Corylus avellana)*	1.14	0.75	0.45	0.23
Cedar	*(Thuja plicata)*		Never injured		
Spruce	*(Picea abies)*		Never injured		
Maple	*(Acer campestre)*		Never injured		

Agricultural species					
Buckwheat	*(Fagopyrum esculentum)*	0.56	0.39	0.26	0.15
Red clover	*(Trifolium pratense)*	0.60	0.50	0.33	0.17
Oats	*(Avena sativa)*	0.63	0.59	0.34	0.17
Peas	*(Pisum sativum)*	0.63	0.59	0.34	0.17
Rhubarb	*(Rheum rhaponticum)*	0.63	0.59	0.34	0.17
Timothy	*(Phleum pratense)*	0.66	0.54	0.40	0.21
Swiss chard	*(Beta vulgaris)*	0.88	0.64	0.42	0.27
Beans	*(Vicia faba)*	0.46	0.45	0.43	0.21
Beets	*(Beta vulgaris)*	1.31	0.77	0.45	0.23
Turnips	*(Brassica rapa)*	1.31	0.77	0.45	0.23
Carrots	*(Daucus carota)*	1.08	0.79	0.50	0.25
Cucumbers	*(Cucumis sativus)*	1.09	0.79	0.50	0.25
Lettuce	*(Lactuca sativa)*	0.64	0.56	0.43	0.38
Radish	*(Raphanus sativus)*	0.64	0.56	0.43	0.38
Squash	*(Cucurbita pepo)*	0.64	0.56	0.43	0.38
Tomatoes	*(Lycopersicon esculentum)*	0.64	0.56	0.43	0.38
Potatoes	*(Solanum tuberosum)*	0.64	0.56	0.43	0.38
Raspberry	*(Rubus idaeus)*	0.74	0.63	0.53	0.39
Celery	*(Apium graveolens)*	0.87	0.74	0.55	0.29
Spinach	*(Spinacia oleracea)*	1.34	0.91	0.50	0.34
Cabbage	*(Brassica oleracea)*	0.94	0.89	0.70	0.45
Corn	*(Zea mays)*		Never injured		

exposure to elevated SO_2 concentrations during the Spring and Summer, when they are metabolically active. Like conifers, deciduous trees also vary greatly with respect to SO_2 sensitivity. London plane (*Platanus* × *hybrida*), red oak (*Quercus borealis*), silver maple (*Acer saccharinum*) and sugar maple (*A. saccharum*) are among the most tolerant deciduous trees, while lodgepole pine (*Pinus contorta*), white spruce (*Picea glauca*) and western red cedar (*Thuja plicata*) are relatively tolerant conifers. Selected planting of these tolerant species may be a temporary solution to deforestation and erosion in denuded areas. The sensitive species, in contrast, may be used as indicator species. Concentrations which result in injury to vegetation are given in Table 9.2.

Relative sensitivities of cultivated plants have also been studied. Many of the grain crops (barley, rye and wheat) are very sensitive to SO_2. Fruit trees are intermediate in sensitivity (apple, peach and apricot) while onions, cucumbers and citrus crops are relatively resistant. Concentrations of SO_2 producing injury for some of these species are also provided in Table 9.2.

The economic impact of SO_2 pollution to crops and forest production is difficult to estimate. Linzon (1971) calculated the cost to the forest industry, as a result of severe foliar injury of eastern white pine in a 1870 km^2 near Sudbury, to be \$117 000 annually where eastern white pine contributed only 7.6% of the total productive forest area. Dreisinger (1964) showed that injury to crops and forests occurred when the average SO_2 concentrations exceeded 0.95, 0.55, 0.35 and 0.25 ppm SO_2 for 1, 2, 4 and 8 h exposure, respectively. Sensitive species were injured at even lower concentrations, while harmful SO_2 fumigations, during a slower growing season, were 4.5 times greater than the threshold limits. According to another estimate (Knabe, 1976), measurable economic effects on agriculture and horticulture can be expected at mean concentrations for the growing season between 0.08 and 0.125 mg SO_2 m^{-3} (0.03 and 0.05 ppm SO_2).

4.2.3 Environmental factors

There are a number of factors which affect the toxicity of SO_2 to vegetation. In general, maximum injury occurs during periods of optimum photosynthesis and metabolic activity. As a result, environmental factors that favour photosynthesis (sunlight, moderate temperatures, high humidity, moderate winds, adequate soil moisture) also favour SO_2 uptake and therefore enhance its toxicity.

Photosynthesis occurs during the day. The stomata are open, enabling rapid exchange of CO_2 and O_2. During the night, the stomata are closed and the plants do not photosynthesize. Consequently, higher SO_2 concentrations can be tolerated by plants during the night than during the day. Similarly, during the winter, when the deciduous trees are dormant and coniferous trees have a reduced metabolic rate, higher SO_2 concentrations can be tolerated without injurious effects. It has been shown, for example, that exposure of sensitive conifers to 2 ppm SO_2 for 69 hr produces no injury in February while the same concentration for only 35 hr can destroy 75% of the foliage in March (Treshow, 1970).

The water content of both the atmosphere (relative humidity) and the soil affects SO_2 toxicity. High relative humidity appears to accelerate the oxidation of SO_2 to SO_3 internally. Externally, SO_2 can be converted into dilute sulphuric acid (H_2SO_4) causing acid burns and lesions. Plants without a protective waxy cuticle are the most sensitive.

Water content in the soil controls the opening of stomata. When the soil moisture is very low, plants wilt and stomata close to prevent water loss. This also has the effect of reducing SO_2 uptake and enables plants to tolerate much higher concentrations. Experiments with alfalfa (*Medicago sativa*) have shown that exposure to 0.75 ppm SO_2 in full sunlight resulted in injury after 11 hr at 40–60% RH (Whitby, 1939).

4.2.4 Threshold concentrations

Based on a large survey of the literature, both the U.S. Department of Health and Welfare and the Research Institute for Public Health Engineering (Netherlands) formulated a set of criteria for atmospheric SO_2 concentrations. There appears to be some discrepancy in their criteria for short-term (<8 hr) and long-term (1 year) exposure, although criteria for intermediate exposure are very similar (Table 9.3). Indeed, for prolonged exposure there is some evidence that no zero effect dose exists (Cooke, 1977).

4.2.5 Interactions with fluoride and ozone

Other gaseous air pollutants can interact synergistically or antagonistically with SO_2 in the atmosphere. It has been suggested that SO_2 in the presence of fluorides might reduce growth indirectly by accelerating leaf abscissions (Treshow, 1970). In contrast, experi-

TABLE 9.3 *Threshold levels of SO_2 for vegetation damage (Linzon, 1972)*

Period of exposure	U.S. Bull. 1619 (1967)		Dutch Bull. G300 (1967)	
(average for)	($\mu g\ m^{-3}$)	(ppm)	($\mu g\ m^{-3}$)	(ppm)
1 hour	1963	0.75	2400	0.92
8 hours	1178	0.45	925	0.35
1 day	785[b]	0.30[a]	700	0.27
1 week	445[b]	0.17[a]	500	0.19
1 month	262	0.10	425	0.16
1 year	79	0.03	390[b]	0.15[b]

[a] Interpolated.
[b] Extrapolated.

ments with SO_2 and O_3, singly and in combination, have shown that the presence of SO_2 reduces O_3 injury in white bean (*Phaseolus vulgaris*) (Beckerson and Hofstra, 1979). This protective action was attributed in part to changes in stomatal response. Since O_3 is a strong oxidizing agent, one would also expect chemical interactions outside the leaf to reduce toxicity of the two gases alone.

4.3 Acid Precipitation and the Aquatic Ecosystem

Many lakes in Canada, Scotland, Scandinavia and the U.S.A. have been altered chemically and biologically as a result of anthropogenic emissions of SO_2 and NO_x into the atmosphere and their subsequent conversion into sulphuric and nitric acid. Those lakes which are sensitive to acidification have soft water, are poorly buffered and occur primarily on hard granitic bedrock where soils are shallow and poorly developed. Very large areas of Canada and Scandinavia occur in such recently glaciated terrain where innumerable lakes have formed. The least sensitive lakes often contain hard water and are well buffered on calcareous bedrock. Changes in these lakes as a result of acid additions will occur more slowly. However, even these well-buffered lakes are threatened if the input of acid continues (Hutchinson and Havas, 1980).

4.3.1 Effects on biota

Perhaps the most important organisms in lakes and streams, from an economical and recreational point of view, are fishes. Significant losses of fish populations were first reported in rivers in southern Norway (Leivestad *et al.*, 1976). The decline of fish could not be attributed to over-fishing or local industrial pollution. The disappearance of fish from lakes and rivers was found to be due both to a failure to reproduce, and to massive fish kills, which occurred in the Spring during the snow melt that coincided with increasingly acidic conditions. Similar results were reported in Canada (Beamish and Harvey, 1972) and in the northeastern U.S.A. (Schofield, 1976). Fish populations in acidifying lakes were found to be ageing, with a high proportion of large, old individuals. Young age classes were rare or absent. Although initially this was very rewarding for fishermen, the fish removed were not replaced naturally, resulting in the eventual extinction of populations. Frequently this happened over a period of a few years. Among the fish, the salmonoids, which are often prized by fishermen, are the most sensitive. Another characteristic of fish in acid lakes is that a high proportion develop deformed backbones, giving them a hunchback appearance (Beamish *et al.*, 1975).

The most critical time of year for aquatic biota is during the Spring when large concentrations of sulphur, nitrogen and H^+ ions, which accumulated during the Winter months in the snow, are rapidly released during snow melt. Many aquatic organisms, which have overwintered in a dormant state, hatch and emerge. Unfortunately, it is often these stages which are most sensitive to acidification. Also, fish and other organisms lay their eggs in shallow waters at this time of acid run-off.

In addition to fish, other aquatic organisms are affected by acidification. Among the algae, blue-greens and greens are very sensitive to low pH (Brock, 1973; Yan and Stokes, 1976). Some diatoms, in contrast, prefer acid water to neutral or alkaline water (Patrick, 1948). A technique used to determine the chemical history of a lake is based on the distribution of diatom species in sediment cores. The assumption is that acid-tolerant diatoms become more common in the sediments as the lake acidifies. Since algae are the primary producers at the base of the food chain in aquatic systems, changes in either species composition or productivity will often trigger changes in the herbivores and carnivores.

Aquatic macrophytes and mosses are also important primary producers, although as a food source for herbivores they play a limited role. One effect of acidification, reported from Sweden and Norway, is the invasion of lake bottoms by *Sphagnum* (Grahn, Hultberg and Landner, 1974). This has a number of serious consequences. *Sphagnum* is a very efficient cation-exchanger. It removes K, Ca, and Mg from the water in exchange for H^+ ions, thereby increasing acidity. Due to its enveloping form of growth, it coats the lake bottom, shutting off nutrient exchange between the sediments and the water column. This exchange is essential for algal growth.

Many aquatic animals such as snails, mussels, clams, crayfish and planktonic crustaceans are very sensitive and are greatly reduced in number by a decrease in pH to 5.0. Sprules (1975) examined zooplankton populations in lakes in Ontario. He found that most species were eliminated once the pH fell to 5.5. Many of these planktonic crustaceans are important food sources for fishes and their decline may contribute to that of fish in acidified lakes.

Insects are much better adapted to low pH than most of the other fauna. Diptera larvae, for example, become the dominant forms as a lake acidifies, possibly as a result of reduced competition for food and other resources (Wiederholm and Eriksson, 1977). Bell (1971) examined the effect of pH on the survival and emergence of nine species of insects. His results showed that caddis fly larvae were most tolerant while mayflies were very sensitive. The 30–day TL (i.e., tolerance limit for 50% mortality within 30 days) was between pH 2.45 and pH 5.38 for all nine species. Successful emergence was much more sensitive to low pH than survival and occurred, on average, one pH unit higher. Therefore, the ability of an organism to survive at a particular pH does not necessarily imply that it will be able to reproduce successfully at that pH.

Another important component of the aquatic ecosystem is decomposers. Their primary contribution is to recycle the organic elements into usable inorganic forms (such as NO_3 and PO_4). In acid lakes the bacteria, which are very sensitive, are replaced by fungi. The fungi are relatively slow decomposers and as they spread along the sediment surface, they tend to reduce nutrient cycling in a fashion similar to that of *Sphagnum*. As a result, acidified lakes are often nutrient-poor. This process has been called "oliogtrophication" which is the opposite of "eutrophication" or nutrient enrichment (Grahn, Hultberg and Landner, 1974). The result of acidification is often a clear lake, with a shift from planktonic organisms to benthic forms. Fish may or may not be present, depending on the pH and on other chemical parameters.

4.3.2 Effects on water chemistry

In addition to the direct toxic effects of H^+ ions to biota, pH also controls a number of chemical processes in lake water. At near neutral pH, the predominant form of carbon is HCO_3^-, which is readily utilized by aquatic plants as a carbon source for photosynthesis. At low pH, HCO_3 is converted into CO_2 which is rapidly lost as it equilibrates with atmospheric CO_2. Those plants that can utilize CO_2, such as *Sphagnum*, are at an advantage in acidified lakes, while other plants may be carbon-limited. If total inorganic carbon is high and the acid input is large during spring run-off, for instance, the CO_2 that is formed can be harmful to fish and other aquatic fauna.

At low pHs, heavy metals present in the soil (and in the lake sediments), for example Al, Fe and Mn become more soluble and are leached into nearby lakes and streams. If the precipitation is acid, but the lake water is still alkaline, the metals will simply precipitate out. However, once the pH of the lake water drops to 4.5 or lower, the metals will remain in solution. This has a number of serious consequences. Aluminium, for example, is extremely toxic to both plants and animals. Threshold toxic concentrations reported for algae, *Daphnia* and fish are between 70 and 1500 $\mu g\ l^{-1}$, respectively (Doudoroff and Katz, 1953; McKee and Wolf, 1963; Biesinger and Christenson, 1972). In Sweden and Norway Gjessing *et al.* (1976) reported elevated Al levels ($\simeq 600\ \mu g\ l^{-1}$) in lake-water at low pH; these concentrations are toxic to some organisms.

Aluminium will also combine with PO_4, to produce insoluble $AlPO_4$, which is unavailable for plant uptake. This is critical since P is an essential nutrient and is often present at low concentrations in natural lake water.

The third major effect of heavy metals at low pH is that some metals act as low pH buffers (Campbell *et al.*, 1964; King *et al.*, 1974). Both Fe and Al, for example, have a strong positive charge which enables them to attract the OH^- group from the water molecule (H_2O) producing free H^+ ions and thereby increasing acidity. The ability of lakes to recover once the acid precipitation stops depends largely on the amount of metals dissolved in the water. Lakes with a high metal concentration will take longer to recover than lakes with a low metal concentration.

4.4 Acid Precipitation and Terrestrial Ecosystems

In contrast to the aquatic ecosystem, significant changes in the terrestrial ecosystem attributed to acid precipitation have not definitely been demonstrated (see Hutchinson and Havas, 1980). This is largely due to the high buffering capacity of the soils compared to the relatively poorly buffered water in the aquatic ecosystem. Consequently, an appropriately longer period of time is necessary for effects to become detectable. A second factor, which may be of significance in protecting terrestrial systems, is that in those geographical areas of North America and Scandinavia where lakes are particularly sensitive to acid inputs, the soils are naturally acidic and the flora and soil fauna have adapted as a result of selection and evolution over thousands of years. Although it is possible that acid precipitation may accelerate natural processes such as leaching, and may affect soil microbial activity, it is probably too early to predict the magnitude and nature of these changes. It is suggested, for example, that in some areas forest growth may be stimulated initially before it declines due to the addition of nitrogen and sulphur in acid rain.

4.4.1 Effects on vegetation

Acid precipitation can have direct effects on vegetation resulting from contact with the foliage, or indirect effects as a result of changes in the soil and micro-organisms inhabiting the soil.

Some of the direct effects are similar to those of SO_2. When the pH of rain is less than 3.0 chlorosis and necrosis may result (Giddings and Galloway 1976; Glass, Glass and Rennie, 1979). This is of particular concern for ornamentals as well as agricultural plants such as spinach, lettuce and tobacco, where the leaves have economic value. Information on the direct effects of acid rain on plants come primarily from simulated acid rain experiments using a wide range of pHs. Effects are observed primarily at extremely low pHs (less than 2.5), which are rarely found in the atmosphere.

In some early experiments conducted by Crowther and Ruston (1911) the effect of rainwater from an industrial city in northern England (Leeds, 1.20 μeq l^{-1} H_2SO_4) and a nearby rural area (Garforth, 2 μeq l^{-1} H_2SO_4 on the growth of Timothy (*Phleum pratense*) was examined over a 3-year period. Additional experiments using neutralized rainwater as well as distilled water adjusted from 2 to 64 μeq l^{-1} H_2SO_4 were also conducted. Their results showed that all plants died within 6 months at 64 μeq l^{-1} and within 1 year at 32 μeq l^{-1}. In Leeds' rainwater and in 64 μeq l^{-1}, germination was reduced and leaves became chlorotic. After 2 years there was a reduced yield in plants exposed to 8 μeq l^{-1} H_2SO_4 or more. Only plants watered with Garforth rain or 2 μeq l^{-1} H_2SO_4 produced normal yields after 3 years. In addition to lower yields, there was a decrease in total plant nitrogen content and increase in crude fibre content at higher H^+ ion concentrations. Therefore, in some cases, yield as well as chemical composition of plants can be significantly altered even at low H_2SO_4 concentrations in rainwater. Interestingly, 70 years after these experiments, interest has been renewed and new studies are under way.

To complicate matters, it has been shown that growth is sometimes stimulated after exposure to acid rain. This stimulation has been attributed to the fertilizer effects of NO_3 and SO_4 (Tveite and Abrahamsen, 1980). Stimulation appears to occur primarily with some coniferous species which have a thick waxy cuticle on their needles and grow on NO_3^- or SO_4^- poor soils.

The effect of acid precipitation on forest productivity is not known. Cogbill (1976) concluded on the basis of tree-ring data that there was no significant change in tree growth that could be attributed to acid precipitation in the northeastern U.S.A. Tree-ring data from Sweden indicated a progressive decrease in forest productivity over the past 15 years (Giddings and Galloway, 1976). This may be attributed to acid precipitation, although it is difficult to isolate acid precipitation as the only mechanism. Annual climatic variability is generally high.

Lichens and mosses are sensitive to acid precipitation and to SO_2 due largely to the fact that they lack a protective waxy cuticle, are not buffered by the soil and obtain their nutrients largely from the atmosphere. Among the lichens, crustose lichens are less sensitive that either fruticose or foliose lichens (Giddings and Galloway, 1976).

In contrast to SO_2, coniferous species are probably more tolerant and better adapted than deciduous species to acid precipitation. This is due partly to the presence of a thick waxy cuticle on evergreen species, and partly to the fact that conifers usually grow on soils which are naturally acidic and, as a result, are adapted to acid conditions. Similar tolerance would be expected for Ericaceous plants, which also have a thick cuticle and grow on acid soils.

The base cations (Ca, Mg, K) are leached more quickly from the foliage by acid rains than by rains near neutrality (Glass, Glass and Rennie, 1979). An excellent review on the effects of precipitation on foliar leaching is given by Tukey (1980). He states that leaching of substances has been implicated in the yield, quality and nutritive value of economic food plants such as grain crops, soft fruits and tropical fruits. The additional impact of H^+ ions in the rain on foliar leaching is not known.

In addition to the direct effects of acid precipitation on plants, there are a number of indirect effects which result from changes in soil chemistry as well as changes in soil microbes. Acidification of soils affects plants in two ways:

(1) it reduces the nutrient status of the soil by leaching nutrients (K, Ca, Mg) from the soils, chemically binding some nutrients (PO_4), as well as by affecting micro-organisms responsible for converting organic forms into usable inorganic forms (NO_3);
(2) it increases soluble levels of metals, such as Al, Fe and Mn, which become phytotoxic at low pH as a result of increased availability. These factors will be discussed in greater detail in the following two sections.

4.4.2 Effects on soil

Soils vary in their sensitivity to acidification. Excellent reviews on this topic are presented in Hutchinson and Havas (1980). Although most soils are well buffered, sandy soils with a low cation exchange capacity (CEC) are perhaps the most sensitive. Neither calcareous soils nor acidic podzols are expected to acidify rapidly as a result of exposure to acid rain (Bache, 1980; Wiklander, 1980).

Even in the balance of observable acidification, exposure to acid rains can have a considerable effect on soils. One documented effect is the leaching of cations such as K, Mg and Ca from soils. The loss of these ions is proportional to the input of H^+ ions into the watershed (Abrahamsen *et al.*, 1976). Odén (1971) predicted that the loss of these elements would result in poorer forest growth. He calculated that approximately 40 years would be necessary for the $\simeq$50% decrease of soil cations in Sweden. In a separate report, Norton (1977) calculated that a significant change in the CEC in areas exposed to acid precipitation could occur between 10 and 100 years. Hutchinson *et al.* (1978) reported substantial changes in the chemistry of soil profiles subjected to intense sulphur acidification in northern Canada.

The solubility of heavy metals, such as Al, Fe and Mn, which are natural constituents of soils, is also affected by acid precipitation. In general, metals become more soluble at low pH and consequently become more available to the biota. Near smelters, the emission of Cu, Ni, Pb, Cd and Zn can also create problems for plant and microbial growth (Tyler, 1974; Jordan and Lechavalier, 1975; Strojan, 1978; Freedman and Hutchinson, 1980). The inability of many plant species to grow on acid soils has been attributed to elevated concentrations of soluble metals (Whitby and Hutchinson 1974; Rorison 1980).

Many of the elements that become mobile eventually are leached into nearby lakes and streams (Johnson, 1979; Hutchinson *et al.*, 1978). Although additions of Ca, Mg and K are not expected to have significant detrimental effects on the aquatic biota, elevated concentrations of heavy metals have been shown to be toxic to these organisms (Doudoroff and Katz, 1953).

4.4.3 Effects on micro-organisms

Soil micro-organisms are essential for the continuous functioning of terrestrial ecosystems. They are necessary in:

(1) converting organic N, P and S into more available inorganic forms;
(2) preventing excessive accumulation of organic matter;
(3) improving soil structure thereby enabling good root development;
(4) degrading toxic substances (Alexander, 1980).

Crowther and Ruston (1911) found that exposure to acid rain resulted in a decrease in the total number of soil bacteria. Nitrogen-fixing and nitrifying organisms were more sensitive than ammonifiers. This resulted in an increase in NH_3 and a decrease in NO_3 and an overall decrease in soil fertility. There is ample evidence that N-fixation decreases in acid soils (Alexander, 1980). In some experiments nitrification was undetectable below pH 4.5. Nodulation involving leguminous plants and *Rhizobium* only occurs above pH 4.6 and nodulation in red clover is poor below pH 5.2. There is also some evidence suggesting that conversion of organic P and S to inorganic forms may also be slower below pH 5 (Alexander, 1980). In acid soils bacteria and actinomycetes are replaced by fungi (Giddings and Galloway, 1976) which are generally slower at decomposing organic matter and recycling nutrients.

Mycorrhizal associations between trees and grass roots and certain fungi are an essential feature of nutrient uptake, especially of P in boreal forest regions. It has been suggested that an increase in soil acidity may reduce the ability of these fungi to infect roots. Infection of legume roots by *Rhizobium* is also at risk.

It has been suggested that plant disease initiation and incidence may be altered in the presence of acid rain. Shriner (1978) examined the effect of simulated rain, pH 3.2 or pH 6.0, on five host-parasite systems. He found that if the acidic rain was applied to plants before inoculation, the severity of the disease increased by 42%. If the acidic rain was applied after inoculation, however, the severity of the disease decreased by 22%. The effects of lower levels of acidity (≃pH 4.0) on host-parasite systems are not known.

4.5 The Impact of Increasing Coal Utilization on the Acid Rain Problem

The discussion with respect to CO_2 applies also in general to the case of acid rain. It is estimated that the total mass of SO_2 emitted from anthropogenic sources is rising annually (Semb, 1978). Granat, Rodhe and Hallberg (1976) estimate that as much as 65% of the annual sulphur emissions are man-made. By the year 2030 a large fraction of global anthropogenic SO_2 emissions will come from coal-fired power stations. There are, however, two special points that need to be made. In the first place, acid rain is a regional phenomenon (as contrasted with CO_2, which is dispersed globally), and the major emitters of SO_2 will generally remain concentrated upwind of sensitive lakes and rivers in Scandinavia and the Laurentian Shield of eastern North America. Secondly, the acid rain problem depends partly on atmospheric chemical transformations involving both SO_2 and NO_x. There is, therefore, need for a broad management control strategy which considers both SO_2 and NO_x emitted from all important types of sources; transportation emissions of NO_x are, for example, concentrated in the same source regions as are the coal-fired power stations.

Whilst it is likely that SO_2 emissions from coal conversion plants will be less than from coal combustion facilities, nitrogen in the air will still be oxidised, resulting in an "acid potential".

5 Trace Element Effects on Ecosystems

5.1 Terrestrial Ecosystems

Several volatile elements in coal pose a potential hazard to the environment but there is little evidence for any marked accumulation in soils around power plants. Similarly, several investigations have failed to detect significant enrichment of trace elements in vegetation or animals in the vicinity of power plants (Roffman, Kary and Hudgins, 1977; Strojan and Turner, 1978). The study of Klein and Russell (1973) is a notable exception, as vegetation as well as soils near a Michigan power plant were reported to contain elevated levels of Cd, Ni and Zn.

The large variability in the trace element content of soils may be masking the input of small quantities of trace elements from coal combustion, not detectable by conventional bulk analysis. The significance of such inputs will be determined to a certain extent by the chemical state of the deposited elements. Greater solubility of an element in fly ash compared to soil will increase its biological availability and potential for bio-accumulation but would also lead to greater mobility, resulting in rapid passage through the soil. Van Hook (1978) considers that microbial soil processes are particularly susceptible to small increments in trace element deposition, as micro-organisms are in intimate contact with trace elements, being found in the immediate vicinity of soil particles. In this context it is of interest to note that reduced litter decomposition rates have been associated with the atmospheric deposition of Cd, Pb and Zn from metal smelters (Tyler, 1976; Coughtrey *et al.*, 1979). Adverse effects on soil/litter processes may also result from the long-term deposition and accumulation of trace elements around power plants. Soils in which ambient levels of trace elements are already elevated are expected to be sensitive to such effects.

A possibility also exists that atmospheric deposition from coal combustion will ameliorate the deficiency of a particular essential trace element in the surrounding soil. Once again, it should be emphasized that the chemical state of the deposited element will determine its biological availability to vegetation. Factors affecting the element's chemical state, and therefore its solubility behaviour with the soil column, include the soil type and its pH, the amount of rain-fall, the acidity of the rainfall and the oxidation-reduction conditions in the soil.

The accumulation of trace elements by terrestrial vegetation growing around ash disposal ponds or on fly ash landfills has been examined recently. Markedly elevated B levels were found in the leaves of several herbaceous species from near one disposal pond (Romney, Wallace and Alexander, 1977). The values encountered would have been high enough to cause toxic symptoms in sensitive species. Yellow sweet clover, (*Melilotus officinalis*) found growing on land-filled fly ash contained higher concentrations of a range of trace elements compared to controls (Furr *et al.*, 1976) but only in the case of Se could those levels be considered elevated. In this respect it is of interest to note that small mammals trapped from this land-fill site also contained greater tissue Se concentrations than controls - the levels present are not, however, considered toxic (Fleming, Gutenmann and Lisk, 1979).

The stabilization and restoration of fly ash disposal areas to reduce water and air erosion has involved the establishment of vegetation on the sites (Hodgson and Buckley, 1975). As part of a feasibility study, Scanlon and Duggan (1979) examined the elemental uptake by eight woody species planted on fly ash. After 2 years' growth there was substantial foliar accumulation in all species of only B, Ni and Se. The foliage of plants grown on fly ash accumulated B to about twice that in plants grown on soil, although levels in the two substrates were similar. Additionally, the Se content of fly ash-grown samples was generally greater than the fly ash concentration itself, yet soil-grown plants contained lower levels than the substrate. These findings suggest that both B and Se display greater bio-availability to plants from fly ash compared to soil. In the case of Ni, the elevated tissue levels of fly ash-grown plants simply reflected the higher fly ash concentrations of this metal compared to soil. After 3 years' plant growth, the rates of element concentration in foliage to content in substrate decreased for both Se and B in the fly ash-grown plants. It would thus appear that there is a reduction in the available B and Se levels in fly ash with time, possibly due to weathering. Reduced availability of B with time has been reported previously for fly ash (Townsend and Gilham, 1975); several years' weathering of fly ash land-fills may sufficiently reduce the B content to allow the growth of susceptible plant species.

The extent of trace element uptake by plants grown on fly ash-amended soil has also been investigated, in order to assess the practical value of fly ash to ameliorate agricultural soil: a range of crops, grown on soil with 5 or 10% fly ash, were found by Furr *et al.* (1976) to contain enhanced concentrations of As, B, Cu, Hg, Ni, Sb and Se in their edible portions, compared to soil-grown controls. In no case, however, were the levels considered elevated or potentially toxic. In a similar study (Furr *et al.*, 1977), cabbage grown on soil amended with fly ash from different regions of the U.S.A. contained concentrations of As, B and Se which were positively correlated with the respective fly ash value. Nevertheless, the levels found in cabbage were much lower than the respective values reported for woody species growing on pure fly ash (Scanlon and Duggan, 1979), even after taking account of the reduced total trace element content of the amended soils. It is possible that the high alkalinity of most fly ashes may actually be responsible for reducing the availability of trace elements from ash when mixed with soil. The finding by Phung *et al.* (1979), that increased fly ash additions to soil actually reduced the available levels of some trace elements, supports this suggestion.

The transfer of potentially toxic elements from plant species grown on fly ash to animals fed such vegetation has recently received some attention (Furr *et al.*, 1976, 1978; Stoewsand, Gutenmann and Lisk, 1978). In all cases, the only element considered elevated in these animals (guinea pig, quail, sheep and goat) was Se, again emphasizing the elements' propensity for bio-accumulation from fly ash. It should be noted, however, that the diets of these animals contained a large proportion of the vegetation grown on pure fly ash – a situation that could be avoided in practice.

Comparison with natural weathering of trace elements

In an attempt to assess the significance of trace element release from solid coal wastes in the U.S.A., Klein *et al.* (1975) compared the quantities of elements discharged annually to ash disposal sites with data on natural weathering mobilization. Similar comparisons have also been carried out on a world-wide basis for Se (Andren, Klein and Talmi, 1975) as well as Cd and Hg (Van Hook, 1978). These studies reported that As, Cu, Ni, Pb and Zn all display mobilization rates from this source that are larger than 10% of the natural weathering rate, while in the case of Cd, Se and U the values are equivalent to those for natural weathering. It is felt, however, that the validity of such comparisons is questionable, as many elements are efficiently contained within the ash disposal sites and show only limited migration into adjacent soil and water systems. In contrast, natural weathering rates reflect the mobilization of trace elements by the world's rivers on a global scale. Nevertheless, in spite of the low mobilization potential of most trace elements from ash disposal sites, future increases in coal utilization will place a greater strain on the sorbing capacity of adjacent soils and may result in greater contamination from such areas. Thus, the extent of mobilization of trace elements from ash disposal sites should be regularly monitored in order to prevent ground water contamination on a significant scale.

5.2 *Aquatic Ecosystems*

The direct run-off of fly ash leachate into adjacent streams has been found to result in the accumulation of several potentially toxic elements in both invertebrate and vertebrate fauna as well as aquatic vegetation (Gutenmann *et al.*, 1976; Cherry and Guthrie, 1977; Cherry *et al.*, 1979). Of particular interest was the elevated concentrations of Se and Cd in several organisms from the drainage system receiving coal ash effluent, examined by Cherry *et al.* (1979). Decreased invertebrate densities adjacent to the ash basin were also reported, but this was ascribed to the increased turbidity, rather than any effect of toxic trace elements. However, in an aquatic bioassay study, Birge (1978) showed that simulated effluent from fly ash was particularly toxic to amphibians and fish. Moreover, LC_{50} values for several trace elements (As, Cd, Hg, Se and Zn) found in coal ash for these organisms were actually lower than the respective drainage water concentrations recorded by Cherry *et al.* (1979). In order to place the extent of this contamination in perspective, it should be noted that Cherry *et al.* (1979) found that ash discharge influence extended to only about 1 km from the disposal pond.

6 Environmental Management

The foregoing sections have emphasized the uncertainties associated with predictions of future emission trends of CO_2, SO_2 and NO_x, and the resulting impacts on climate, the biosphere and society. There is evidently need for further research and this will grow quite naturally in response to emerging priorities set by governments and intergovernmental bodies. But there is also urgent need for engineering order-of-magnitude assessments of some of these great global problems. The assessments should provide decision-makers with alternative management strategies, even though there is still great uncertainty. Here it should be recognized that because environmental management actions usually change the *status quo* only incrementally, mitigation measures must be viewed over the long term, and they must be initiated as quickly as possible.

In the sense used here, an assessment is *not* a research project but is a management tool. The elements of an international environmental assessment should therefore include the following (Munn, 1979):

(1) A brief description of relevant past and present biogeophysical and socioeconomic states and trends, separating natural from anthropogenic components.
(2) A brief description of present biogeophysical and socioeconomic

impacts, including identification of the affected parties (who is gaining and who is losing?).

(3) A study of several alternative management strategies, including that of non-intervention; for each alternative:
  (a) estimates of the biogeophysical changes and socioeconomic benefits and disbenefits, both over the short and long term, clearly identifying the affected parties;
  (b) a sensitivity analysis, providing estimates of the uncertainty ranges in the predictions;
  (c) estimates of the possibilities and costs of mitigating measures.

(4) A comparison of alternatives.

(5) A non-technical summary.

A problem with many action-forcing plans for mitigating environmental impacts is that "research needs" and "control strategies" are in adversary positions, whereas both are essential components of reasonable environmental management programmes.

In relation to the environmental impacts of increasing coal utilization, component (1) above and part of component (2) have been covered in this chapter. To go further has been a goal of international programmes in Europe (OECD, ECE) and North America. The industrial and power-generation sectors have, in general, been willing to do their fair share in reducing emissions. But what is a fair share - a percentage reduction in emissions, a percentage increase in costs to reduce emissions, or a percentage reduction in the levels of exposures of receptors? In addition, should present imbalances in emissions be a starting point for control strategies? These are difficult questions to answer, but they should be examined within component (3) above of an international environmental assessment.

One final recommendation. A major goal of environmental management should be to keep as many options open as possible, thus reducing the chances of surprise and of being unable to cope with unexpected biogeophysical and economic impacts.

7 References

ABRAHAMSEN, G., BJOR, K., HORNTVEDT, R. and TVEITE, B. (1976). Effects of acid precipitation on coniferous forests. *Impacts of Acid Precipitation on Forest and Freshwater Ecosystems in Norway* (ed. by F. H. Braekke), pp. 37–63. SNSF Project, Aas-NLH, Norway.

ALEXANDER, M. (1980). Effects of acidity on micro-organisms and microbial processes in soil. *Effects of Acid Precipitation on Terrestrial Ecosystems* (ed. by

T. C. Hutchinson and M. Havas), pp. 363–74. Plenum Press, New York.

ANDREN, A. W., KLEIN, D. H. and TALMI, Y. (1975). Selenium in coal-fired steam plant emissions. *Environmental Science and Technology*, **9**, 856–8.

BACHE, B. W. (1980). The sensitivity of soils to acidification. *Effects of Acid Precipitation on Terrestrial Ecosystems* (ed. by T. C. Hutchinson and M. Havas), pp. 569–72. Plenum Press, New York.

BEAMISH, R. J. and HARVEY, H. H. (1972). Acidification of the La Cloche Mountain Lakes, Ontario, and resulting fish mortalities. *Journal of the Fisheries Research Board of Canada*, **29**, 1131–43.

BEAMISH, R. J., LOCKHART, W. L., VAN LOON, J. C. and HARVEY, H. H. (1975). Long-term acidification of a lake and resulting effects on fishes. *Ambio*, **4**, 90–102.

BECKERSON, D. W. and HOFSTRA, G. (1979). Stomatal responses of white bean to O_3 and SO_2 singly or in combination. *Atmospheric Environment*, **113**, 533–5.

BELL, H. L. (1971). Effect of low pH on the survival and emergence of aquatic insects. *Water Research*, **5**, 113–19.

BELL, J. N. B., RUTTER, A. J. and RELTON, J. (1979). Studies on the effects of low levels of sulphur dioxide on the growth of *Lolium perenne* L. *New Phytologist*, **83**, 627–43.

BIESINGER, K. E. and CHRISTENSON, G. M. (1972). Effects of various metals on survival, growth, reproduction and metabolism of *Daphnia magna*. *Journal of the Fisheries Board of Canada*, **29**, 1691–700.

BIRGE, W. J. (1978). Aquatic toxicology of trace elements of coal and fly ash. *Energy and Environmental Stress in Aquatic Systems* (ed. by J. H. Thorp and J. W. Gibbons), pp. 219–40. Symposium Augusta, Georgia 1977. U.S. Department of Energy.

BOLIN, B. (ed.) (1980). *Modelling the Global Carbon Cycle*. Wiley, New York.

BROCK, T. D. (1973). Lower pH limit for the existence of blue-green algae: evolutionary and ecological implications. *Science*, **179**, 480–3.

BRYSON, R. A. and GOODMAN, B. M. (1980). Volcanic activity and climatic change. *Science*, **207**, 1041–4.

CAMPBELL, R. S., LIND, O. T., GEILING, W. T. and HARP, G. L. (1964). *Recovery from Acid Pollution in Shallow Strip-mine Lakes in Missouri*. Proceedings of the 19th Industrial Waste Conference, Purdue University, Lafayette, Indiana, Engineering Extension Series no. 117.

Chemical Engineering (1977). 3 January, p. 79. McGraw-Hill, New York.

CHERRY. D. S. and GUTHRIE, R. K. (1977). Toxic metals in surface waters from coal ash. *Water Resources Bulletin*, **13**, 1227–36.

CHERRY, D. S., GUTHRIE, R. K., SHERBERGER, F. F. and LARRICK, S. R. (1979). The influence of coal ash and thermal discharges upon the distribution and bioaccumulation of aquatic invertebrates. *Hydrobiologia*, **62**, 257–67.

COGBILL, C. V. (1976). The effect of acid precipitation on tree growth in eastern North America. *Proceedings of the 1st International Symposium on Acid Precipitation and the Forest Ecosystem*, pp. 1027–32. USDA Forest Service General Technical Report NE–23.

COOKE, N. E. (ed.) (1977). *Sulphur and its Inorganic Derivatives in the Canadian Environment*. Publication no. NRCC 15015, National Research Council of Canada, Ottawa.

COUGHTREY, P. J., JONES, C. H., MARTIN, M. H. and SHALES, S. W. (1979). Litter accumulation in woodlands contaminated by Pb, Zn, Cd and Cu. *Oecologia* (Berl.) **39**, 51–60.

Crowther, C. and Ruston, A. G. (1911). The nature, distribution and effects upon vegetation of atmospheric impurities in and near an industrial town. *Journal of Agricultural Science,* **4**, 25–55.

DOE (1980). *CO_2 Research Progress Report.* CO_2 and Climate Research Program, Department of Energy, Washington, DC., 130pp.

Doudoroff, P. and Katz, M. (1953). Critical review of literature on the toxicity of industrial wastes and their components to fish II. Metals, as salts. *Sewage and Industrial Wastes,* **25**, 802–39.

Dreisinger, B. R. (1964). *The Impact of Sulphur Dioxide Pollution of Crops and Forests.* Background Paper A–4–2–1– for Canadian Council of Resource Ministries.

Emmanuel, W. R., Killough, G. G. and Olson, J. S. (1980). Modelling the circulation of carbon in the world's terrestrial ecosystems. *Modelling the Global Carbon Cycle* (ed. by B. Bolin). Wiley, New York.

Engström, A. (1971). *Air Pollution Across National Boundaries: the Impact on the Environment of Sulfur in Air and Precipitation.* Sweden's Case Study for the U.N. Conference on the Human Environment.

Eriksson, E. (1963). The yearly circulation of sulfur in nature. *Journal of Geophysics Research*, **68**, 4001–8.

Fleming, W. J., Gutenmann, W. H. and Lisk, D. J. (1979). Selenium in tissues of woodchucks inhabiting contaminated and landfills. *Bulletin of Environmental Contamination and Toxicology* **21**, 1–3.

Freedman, B. and Hutchinson, T. C. (1980). Smelter pollution near Sudbury, Ontario, Canada, and effects on forest litter decomposition. *Effects of Acid Precipitation on Terrestrial Ecosystems* (ed. by T. C. Hutchinson and M. Havas), pp. 359–434. Plenum Press, New York.

Furr, A. K., Kelly, W. C., Bache, C. A., Gutenmann, W. H. and Lisk, D. J. (1976). Multielement uptake by vegetables and millet grown in pots on fly ash amended soil. *Journal of Agricultural Food Chemistry,* **24**, 885–8.

Furr, A. K., Parkinson, T. F., Hinrichs, R. A., Van Campen, D. R., Bache, C. A., Gutenmann, W. H., St John, L. E., Pakkala, I. S. and Lisk, D. J. (1977). National survey of elements and radioactivity in fly ashes: absorption of elements by cabbage grown in fly ash soil mixtures. *Environmental Science and Technology*, **11**, 1194–201.

Furr, A. K., Parkinson, T. F., Heffron, C. L., Reid, J. T., Haschek, W. M., Gutenmann, W. H., Bache, C. A., St John, L. E. and Lisk, D. J. (1978). Elemental content of tissues and excreta of lambs, goats and kids fed white sweet clover growing on fly ash. *Journal of Agricultural Food Chemistry,* **26**, 847–51.

Giddings, J. and Galloway, J. N. (1976). *The Effects of Acid Precipitation on Aquatic and Terrestrial Ecosystems.* EEP–2, Cornell University, Ithaca.

Gilbert, O. L. (1965). Lichens as indicators of air pollution in the Tyne Valley. *Ecology and the Industrial Society* (ed. by G. T. Goodman, R. E. Edwards and J. M. Lambert), pp. 35–47. Wiley, New York.

Gjessing, E. T., Henriksen, A., Johannessen, M. and Wright, R. F. (1976). Effects of Acid Precipitation on Freshwater Chemistry. *Impact of Acid Precipitation of Forest and Freshwater Ecosystems in Norway* (ed. by F. H. Braekke). Research Report 6/76, SNSF Project, Norway.

Glass, N. R., Glass, G. E. and Rennie, P. J. (1979). Effects of acid precipitation. *Environmental Science and Technology,* **13**, 1350–5.

Grahn, O., Hultberg, H. and Landner, L. (1974). Oligotrophication - a self-accelerating process in lakes subjected to excessive supply of acid substances. *Ambio*, **3**, 93–4.

GRANAT, L., RODHE, H. and HALLBERG, R. O. (1976). The global sulphur cycle. *Nitrogen, Phosphorus and Sulphur - Global Cycles* (ed. by B. H. Svenson and R. Söderlund). Ecological Bulletin 22, Stockholm.

GUTENMANN, W. H., BACH, C. A., YOUNGS, W. D. AND LISK, D. J. (1976). Selenium in fly ash. *Science*, **191**, 966–7.

HODGSON, D. R. and BUCKLEY, G. P. (1975). A practical approach towards establishment of trees and shrubs on pulverized fuel ash. *The Ecology of Resource Degradation and Renewal* (ed. by M. J. Chadwick and G. T. Goodman), pp. 305–29, Blackwell, Oxford.

HUTCHINSON, T. C. and HAVAS, M. (eds.) (1980). *Effects of Acid Precipitation on Terrestrial Ecosystems.* Plenum Press, New York.

HUTCHINSON, T. C. and WHITBY, L. M. (1974). Heavy metal pollution in the Sudbury mining and smelting region of Canada. I. Soil and vegetation contamination by nickel, copper and other metals. *Environmental Conservation*, **1**, 123–32.

HUTCHINSON, T. C., GIZYN, W., HAVAS, M. and ZOBENS, V. (1978). Effect of long-term lignite burns on arctic ecosystems at the Smoking Hills, N.W.T. *Trace Substances in Environmental Health*, XII (ed. by D. D. Hemphill), pp. 317–32. University of Missouri, Columbia.

IDSO, S. B. (1980). The climatological significance of a doubling of earth's atmospheric CO_2 concentration. *Science*, **207**, 1462–3.

IMBRIE, J. and IMBRIE, J. Z. (1980). Modelling the climatic response to orbital variations. *Science*, **207**, 943–53.

JEFFREE, C. E. (1979). Plant damage caused by SO_2. *Proceedings of the Symposium on the Effects of Airborne Pollution on Vegetation.* United Nations Economic Commission for Europe.

JOHNSON, A. H. (1979). Acidification of headwater streams in the New Jersey pinebarrens. *Journal of Environmental Quality*, **8**, 383–6.

JORDAN, M. J. and LECHAVALIER, M. P. (1975). Effects of zinc smelter emissions on forest soil micro flora. *Canadian Journal of Microbiology*, **21**, 1855.

KATZ, M. (ed.) (1939). *Effects of Sulphur Dioxide on Vegetation.* National Research Council (Canada) Publication No. 815, 447 pp.

KEELING, C. D. (1973). Industrial production of CO_2 from fossil fuels and limestone. *Tellus*, **25**, 174.

KEELING, C. D. and BACASTOW, R. B. (1977). Impacts of industrial gases on climate. *Energy and Climate.* U. S. National Academy of Science, Washington, DC, pp. 72–95.

KING, D. L., SIMMLER, J. J., DECKER, C. S. and OGG, C. W. (1974). Acid strip mine lake recovery. *Journal of the Water Pollution Control Federation*, **46**, 2301–15.

KLEIN, D. H. and RUSSELL, P. (1973). Heavy metals: fallout around a power plant. *Environmental Science and Technology*, **7**, 357–8.

KLEIN, D. H., ANDREN, A. W., CARTER, D. A., EMERY, D. F., FELDMAN, C., FULKERSON, W., LYON, W. S., OGLE, D. C., TALMI, Y., VAN HOOK, R. I. and BOLTON, N. (1975). Pathways of thirty-seven trace elements through coal-fired power plant. *Environmental Science and Technology*, **9**, 937–9.

KNABE, W. (1976). Effects of sulphur dioxide on terrestrial vegetation. *Ambio*, **5**, 213–18.

LeBlanc, F. and DeSloover, J. (1970). Relation between industrialisation and the distribution and growth of epiphytic lichens and mosses in Montreal. *Canadian Journal of Botany,* **48**, 1485–96.

Leivestad, H., Hendrey, G., Muniz, I. P. and Snekvik, E. (1976). Effects of acid precipitation on freshwater organisms. *Impact of Acid Precipitation on Forest and Freshwater Ecosystems in Norway* (ed. by F. H. Braekke). Research Report 6/76, SNSF Project, Aas–NLH, Norway.

Likens, G. E., Wright, R. F., Galloway, J. N. and Butler, T. J. (1979). Acid rain. *Scientific American,* **241**, October, 43–51.

Linzon, S. N. (1971). Economic effects of sulphur dioxide on forest growth. *Journal of the Air Pollution Control Association,* **21**, 81–6.

Linzon, S. N. (1972). Effects of sulphur oxides on vegetation. *The Forestry Chronicle,* pp. 182–6.

MacFarland, H. N. (ed.) (1975). *Photochemical Air Pollution: Formation, Transport and Effects.* Publication no. NRCC 14096, National Research Council of Canada, Ottawa.

Manabe, S. and Stouffer, R. J. (1979). A CO_2-climate sensitivity study with a mathematical model of the global climate. *Nature,* **282**, 491–6.

Manabe, S. and Wetherald, R. T. (1980). On the distribution of climate change resulting from an increase in CO_2 content of the atmosphere. *Journal of Atmospheric Sciences,* **37**, 99–118.

McKee, J. E. and Wolf, H. W. (eds.) (1963). *Water Quality Criteria* (2nd ed). California State Resources Control Board, Publication 3–A.

Munn, R. E. (1973). Secular increases in summer haziness in the Atlantic provinces. *Atmosphere,* **11**, 156–61.

Munn, R. E. (1979). The framework for a climate impact assessment. *CO_2 Issues and Impacts,* pp. 19–22. Atmospheric Environment Service, 4905 Dufferin St., Downsview, Ontario, Canada.

Munn, R. E. and Machta, L. (1979). Human activities that affect climate. *Proceedings of the World Climate Conference,* pp. 170–209. WMO, Geneva.

Newell, R. E. and Dopplick, T. G. (1979). Questions concerning the possible influence of anthropogenic CO_2 on atmospheric temperature. *Journal of Applied Meteorology,* **18**, 822–5.

Norton, S. A. (1977). Changes in chemical processes in sails caused by acid rain. *Water, Air and Soil Pollution,* **7**, 389–400.

Novakova, E. (1969). Influence des pollutions industrielles sur les communautés animales et l'utilisation des animaux comme bioindicateurs. *Air Pollution,* pp. 41–50. PUDOC, Wageningen.

Novakova, E. and Paukert, J. (1980). Schwankungne des Schwefelgehaltes im Harn des Feldhasen als indikator für Umweltverunreinigungen. Proc. 3rd. International Conference, *Bioindicatores Deteriorisationis Regionis,* pp. 289–94. Academia, Praha.

Odén, S. (1971). Acidification by atmospheric precipitation — a general threat to the ecosystem. *Acidification of the Biological Environment* (ed. by I. Mysterud). Journal of the Fisheries Research Board of Canada, Translation, Series No. 2564.

Patrick, R. (1948). Factors affecting the distribution of diatoms. *Botanical Revue,* **14**, 437–524.

Phung, H. T., Lund, L. J., Page, A. L. and Bradford, G. R. (1979). Trace elements in fly ash and their release in water and treated soils. *Journal of*

Environmental Quality, **8**, 171–5.

RAO, D. N. and LEBLANC, F. (1966). Effects of sulphur dioxide on the lichen alga, with special reference to chlorophyll. *Bryologist*, **69**, 69–75.

RENZETTI, N. A. and DOYLE, G. J. (1960). Photochemical aerosol formation in sulphur dioxide-hydrocarbon systems. *International Journal of Air Pollution*, **2**, 327–45.

ROBINSON, E. and ROBBINS, R. A. (1968). Sources, abundance and fate of gaseous atmospheric pollutants. Rep. Project SCC–8501, Stanford Research Institute, Menlo Park, California, U.S.A.

RODHE, H., CRUTZEN, P. and VANDERPOL, A. (1979). Formation of sulphuric and nitric acid in the atmosphere during long-range transport. *Proceedings of the Symposium on Long-Range Transport of Pollutants*, pp. 165–71. Sofia, Bulgaria, WMO, Geneva.

ROFFMAN, H. K., KARY, R. E. and HUDGINS, T. (1977). Ecological distribution of trace elements emitted by coal-burning power generating units employing scrubbers and electrostatic precipitators. *Fourth Symposium on Coal Utilization*, pp. 192–215. Louisville, Kentucky.

ROMNEY, E. M., WALLACE, A. and ALEXANDER, G. V. (1977). Boron in vegetation in relation to a coal-burning power plant. *Communications in Soil Science and Plant Analysis*, **8**, 803–7.

RORISON, I. H. (1980). The effects of soil acidity on nutrient availability and plant response. *Effects of Acid Precipitation on Terrestrial Ecosystems* (ed. by T. C. Hutchinson and M. Havas), pp. 283–304. Plenum Press, New York.

ROTTY, R. M. (1979). Present and future production of CO_2 from fossil fuels - a global appraisal. *Workshop on the Global Effects of CO_2 from Fossil Fuels*. U.S. Department of Energy Conference 770385–CU–11.

SANDSTRÖM, K–R. and HÄLLGREN, J. E. (1973). Using lichens as physiological indicators of sulfurous pollutants. *Ambio*, **2**, 13–21.

SCANLON, D. H. and DUGGAN, J. C. (1979). Growth and element uptake of woody plants on fly ash. *Environmental Science and Technology*, **13**, 311–15.

SCHOFIELD, C. L. (1976). Acid precipitation: effects on fish. *Ambio*, **5**, 228–30.

SCOPE (1979). *The Global Carbon Cycle* (ed. by B. Bolin, E. T. Degens, S. Kempe and P. Ketner). Wiley, New York.

SEMB, A. (1978). Sulphur emissions in Europe. *Sulphur in the Atmosphere* (ed. by R. B. Husan, J. P. Lodge and D. J. Moore). Pergamon, Oxford.

SHRINER, D. S. (1978). Effects of simulated acidic rain on host parasite interactions in plant diseases. *Phytopathology*, **68**, 213–18.

SIDEBOTTOM, H. W., BADCOCK, C. C., JACKSON, G. E., CALVERT, J. G., REINHARDT, G. W. and DAMON, E. K. (1972). Photooxidation of sulphur dioxide. *Environmental Science and Technology*, **6**, 72–8.

SPRULES, W. G. (1975). Midsummer crustacean zooplankton communities in acid-stressed lakes. *Journal of the Fisheries Research Board of Canada*, **32**, 389–95.

STEINBERG, M. and ALBANESE, A. S. (1980). Environmental control technology for atmospheric CO_2. *Proceedings of the International Workshop on Energy/ Climate Interaction*, Münster, F.R.G.

STOEWSAND, G. S., GUTENMANN, W. H. and LISK, D. J. (1978). Wheat grown on fly ash: high selenium uptake and response when fed to Japanese quail. *Journal of Agricultural Food Chemistry*, **26**, 757–9.

STROJAN, C. L. (1978). Forest litter decomposition in the vicinity of a zinc smelter. *Oecologia*, **32**, 203–12.

STROJAN, C. L. and TURNER, F. B. (1978). Trace elements and sulfur in soils and plants near the Mohave Generating Station in Southern Nevada. *4th Joint Conference on Sensing of Environmental Pollutants*, pp. 537–40. American Chemical Society, Washington, DC.

TOWNSEND, W. N. and GILHAM, E. W. F. (1975). Pulverized fuel ash as a medium for plant growth. *The Ecology of Resource Degradation and Renewal* (ed. by M. J. Chadwick and G. T. Goodman), pp. 287–304. Blackwell, Oxford.

TRESHOW, M. (1970). *Environment and Plant Response*. McGraw-Hill, New York.

TUKEY, H. B., JR (1980). Some effects of rain and mist on plants, with implications for acid precipitation. *Effects of Acid Precipitation on Terrestrial Ecosystems*. (ed. by T. C. Hutchinson and M. Havas), pp. 141–50. Plenum Press, New York.

TVEITE, B. and ABRAHAMSEN (1980). Effects of artificial acid rain on the growth and nutrient status of trees. *Effects of Acid Precipitation on Terrestrial Ecosystems* (ed. by T. C. Hutchinson and M. Havas), pp. 305–18. Plenum Press, New York.

TYLER, G. (1974). Heavy metal pollution and soil enzymatic activity. *Plant and Soil*, **41**, 303–11.

TYLER, G. (1976). Heavy metal pollution, phosphatase activity and mineralization of organic phosphorus in forest soils. *Soil Biology and Biochemistry*, **8**, 327–32.

U.S.A./CANADA (1979). *The LRTAP Problem in North America: a Preliminary Overview*. U.S. - Canada Research Consultation Group on LRTAP. State Department, Washington and External Affairs, Ottawa.

VAN HOOK, R. I. (1978). Potential health and environmental effects of trace elements and radionuclides from increased coal utilization. Report No. ORNL–5367. Oak Ridge National Laboratory, Tennessee.

WALLEN, C. C. (1980). A preliminary evaluation of BAPM on precipitation chemistry data for Europe and North America during 1972–6. *WMO Technical Conference on Regional and Global Observation of Atmospheric Pollution*. Boulder, Colorado. WMO, Geneva.

WELLBURN, A. R., MAJERNIK, O. and WELLBURN, F. A. M. (1972). Effects of SO_2 and NO_2 polluted air upon the ultrastructure of chloroplasts. *Environmental Pollution*, **3**, 37–49.

WHITBY, G. S. (1939). The effects of sulphur dioxide on vegetation. *Chemistry and Industry*, **11**, 991–9.

WHITBY, L. M. and HUTCHINSON, T. C. (1974). Heavy-metal pollution in the Sudbury mining and smelting region of Canada II. Soil toxicity tests. *Environmental Conservation*, **1**, 191–200.

WIEDERHOLM, T. and ERIKSSON, L. (1977). Benthos of an acid lake. *Oikos*, **29**, 261–7.

WIKLANDER, L. (1980). The sensitivity of soils to acid precipitation. *Effects of Acid Precipitation on Terrestrial Ecosystems* (ed. by T. C. Hutchinson and M. Havas), pp. 553–67. Plenum Press, New York.

WILLIAMS, J. (ed.) (1979). *Carbon Dioxide, Climate and Society*. Pergamon Press, New York.

WOODWELL, G. M., MACDONALD, A. J., REVILLE, R. and KEELING, C. D. (1979). *The Carbon Dioxide Problem: Implications for Policy in the Management of*

Energy and Other Resources. Report to the President's Council on Environmental Quality, Washington, DC.
YAN, N. D. and STOKES, P. M. (1976). The effects of pH on lake water chemistry and phytoplankton in a La Cloche mountain lake. *Proceedings 11th Canadian Symposium on Water Pollution Research in Canada*, pp. 127–37.

Index